# Lecture Notes in Mathematics

Edited by A. Dold and B. Eckmann

414

T. Kambayashi
M. Miyanishi
M. Takeuchi

# Unipotent Algebraic Groups

Springer-Verlag
Berlin · Heidelberg · New York 1974

Prof. Dr. Tatsuji Kambayashi
Department of Mathematics
Northern Illinois University
DeKalb, IL 60115/USA

Prof. Dr. Masayoshi Miyanishi
Department of Mathematics
Osaka University
Toyonaka, Osaka 560/Japan

Prof. Dr. Mitsuhiro Takeuchi
Department of Mathematics
University of Tsukuba
Sakura-mura, Niihari-gun
Ibaraki-ken 300-31/Japan

**Library of Congress Cataloging in Publication Data**

Kambayashi, Tatsuji, 1933-
Unipotent algebraic groups.

(Lecture notes in mathematics ; 414)
Bibliography: p.
Includes index.
1. Linear algebraic groups. 2. Group schemes (Mathematics) 3. Commutative rings. I. Miyanishi, Masayoshi, 1940- joint author. II. Takeuchi, Mitsuhiro, 1947- joint author. III. Title. IV. Series: Lecture notes in mathematics (Berlin) ; 414.
QA3.L28 no. 414 [QA171] 510'.8s [512'.2] 74-20780

AMS Subject Classifications (1970): 13B10, 13D15, 13F15, 13F20
14G05, 14L15, 16A24, 20G15

ISBN 3-540-06960-7 Springer-Verlag Berlin · Heidelberg · New York
ISBN 0-387-06960-7 Springer-Verlag New York · Heidelberg · Berlin



Offsetdruck: Julius Beltz, Hemsbach/Bergstr.

# PREFACE

The geometry and group theory of unipotent algebraic groups over an arbitrary ground field were successfully pioneered by Rosenlicht in the late fifties and early sixties. In the subsequent years not very much was added to the knowledge in this area, with only a few notable contributions such as those by Russell and by Tits. Lately, however, there have been indications of growing interest in this and related subject areas (affine space, its automorphisms, purely inseparable cohomology theories, ... ). Even as the present paper was undergoing the final redaction, a graduate student in Tokyo settled our conjecture in Section 5 by constructing an elaborate counter-example; one of the coauthors established the absence of nontrivial separable forms of the affine plane, confirming an earlier announcement of Shafarevich; another found a description of the category of all commutative affine group schemes over an imperfect field by extending Schoeller's work; and still another obtained an algebraic characterization of the affine plane.

The material presented here might be made into two or three separate research papers of a more polished character. Instead, in view of the rapid developments as indicated above and because of our belief in the unity behind our work, we have chosen to publish our results as one whole and as quickly as possible. We are thankful to the editors and the publishers of the Lecture Notes series for providing us with an ideal outlet for our joint work. It is our sincere hope that this publication will serve to stimulate further research in this field full of deep and fascinating problems.

Finally, our grateful acknowledgements are due to the Research Institute for Mathematical Sciences, Kyoto University for the hospitality extended to one of us while the research for the present paper was conducted during the year 1972-73; to the young ladies on the Institute's staff for the carefull and efficient typing of the manuscript; and to the National Science Foundation for partially supporting the final preparation of the manuscript through a research grant.

June 1974

The Coauthors

TABLE OF CONTENTS

# LEITFADEN

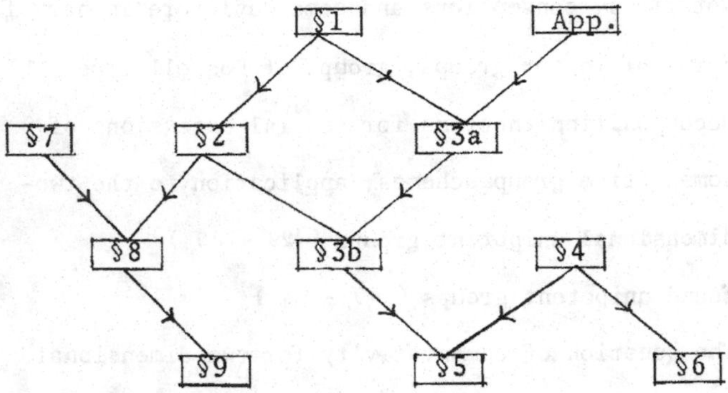


§3a = 3.1 to 3.4;  §3b = 3.5 to 3.7

# On the theory of unipotent algebraic groups over an arbitrary ground field

by

Tatsuji Kambayashi   Masayoshi Miyanishi   and Mitsuhiro Takeuchi

## Introduction

This paper reports on our joint investigation of both the group-theoretical and the geometric structures of unipotent algebraic groups defined over an arbitrary ground field. Let G be such a group, connected and defined over a field k. In case k is perfect, the underlying variety $\bar{G}$ of G is known to be k-isomorphic to the affine space $\mathbb{A}^n$ of dimension n = dim G (cf. §8 and Appendix). It is also known that, conversely, if $\mathbb{A}^n$ is given a structure of algebraic group over quite an arbitrary field then the resulting algebraic group is unipotent — a fact due to Lazard (cf. §8). Thus, the geometric structure of a unipotent algebraic group is completely known over a perfect ground field, while over an imperfect field the study of the structure leads to the study of forms of $\mathbb{A}^n$. As for the group structure of a connected unipotent algebraic group G, one knows that if the ground field $k$ is perfect $G$ possesses a central series of k-closed subgroups in which every successive quotient is k-isomorphic to the one-dimensional vector group $G_a$ (cf. §1 and §8). Beyond that, practically nothing is known even over a perfect $k$, except when $\dim G \leq 2$: $G$ is k-isomorphic to $G_a$ if $\dim G = 1$, and at dimension 2 the central extensions of $G_a$ by $G_a$ has been completely calculated, so that one knows all such $G$ with $\dim G = 2$ (cf. §3). The state of our knowledge in this regard gets even worse when $k$ is imperfect: Russell has determined all connected one-dimensional unipotent groups, extending Rosenlicht's earlier discovery of nontrivial forms of $G_a$; but at dimension >1 little else has been known.

We have extended the results summarized above in various directions, and have also started some new lines of investigation. The contents of the present paper will now be explained: Very roughly, Part I is concerned with the group structure of the unipotent group, while Part II does the geometry of such group. In more detail: Extending Russell's work, we determine in §2 all k-forms of the vector group $G_a \times \cdots \times G_a$ (n copies, $n \geq 1$). To a nontrivial k-form of $G_a$ expressed in terms of a standardized two-indeterminate equation we have given the name of a "k-group of Russell type". In §3 we study the structure of $\mathrm{Ext}_{\mathrm{cent}}(B,A)$, the equivalence classes of central extensions of commutative k-group schemes A, B.

After proving a general decomposition theorem of $\mathrm{Ext}_{\mathrm{cent}}(B,A)$, we make the theorem more precise in the special case when both A and B correspond to k[F]-modules. It will be shown how this theorem together with the results of §2 yield a sort of classification of all two-dimensional unipotent k-groups. In §4 we introduce after Tits the important concept of "k-wound unipotent k-group" and give a new proof to his basic result on the concept. Section 5 (§5) is concerned mainly with the conjecture that every k-wound unipotent k-group of dimension 2 is commutative. We present a counter-example to the truth of the conjecture, but we also present a number of results on the central extensions between k-groups of Russell type, which will show the plausibility of the conjecture. Section 6 (§6) ultimately aims at determining all k-forms of $\mathbb{A}^1$. We fall short of the goal, but we have determined all such forms that are k-rational, of genus 0 or of genus 1. We have also calculated the Picard groups of the underlying schemes of certain Russell type k-groups. In §7 we give a criterion in terms of Demazure-Hochschild cohomology in order for the action of a unipotent algebraic k-group scheme on an affine k-scheme to have an affine representable quotient. Using the criterion in part, we characterize in §8 the underlying scheme of a unipotent k-group both in terms of k-forms of vector groups and in terms of the make-up of the affine algebra of its underlying k-scheme. The last section

(§9) gives a characterization of unipotency of an affine k-group scheme in terms of its hyperalgebra as introduced by one of the authors. In the Appendix which goes over the generalities on central extensions and proves a six-term exact sequence, we also discuss the splitting of extensions with kernel $G_a$ and some applications.

Some of our results are valid over arbitrary ground fields. Others lose their significance if the ground field has characteristic 0. In the main, our theory is of interest over an imperfect ground field.

Important previous contributions on the subject of the present article include three pioneering papers of Rosenlicht [8], [9], [10], Russell's paper [11] and Tits' lecture notes [15; esp. Chap. IV, §4]. For reference material, we have made a heavy use of DG and SGAD (see References at the end for the abbreviations).

Our notations and conventions are rather conformist, and are explained in §1 below. Let us reiterate here only that our "k-group" is synonymous with "k-smooth k-group scheme".

# Part 1

## 1. Notations, conventions and some basic preliminery facts

1.0. Notaions and conventions. Throughout, k denotes a field of arbitrary characteristic. The letter p is reserved to represent the characteristic of k when this is a prime number. The algebraic closure and the separable closure of k are respectively denoted by $\bar{k}$ and $k_s$. The reference to the ground field or the ground scheme will be usually omitted if it is respectively to k or to Spec k. Thus, e.g., an algebra means k-algebra and $\otimes$ stands for the tensor product $\otimes_k$ over k.

Categories are denoted by sanserif letters. Principal among them are the following: $\mathsf{Alg}_k$ := the category of all commutative, unital k-algebras; $\mathsf{E}$ := the category of all sets; $\mathsf{Gr}$ := the category of all groups; $\mathsf{Ab}$ := the category of all abelian groups; $\mathsf{Sch}_k$ := the category of all k-schemes; $\mathsf{Aff.Sch}_k$ := the category of all affine k-schemes. If $\mathsf{C}$, $\mathsf{D}$ are categories, $\mathsf{CD}$ will mean the category of all functors from $\mathsf{C}$ to $\mathsf{D}$. Thus, for instance, $\mathsf{Alg}_k\mathsf{Gr}$ is the category of all k-group functors.

The letter R will be used exclusively for the general object in $\mathsf{Alg}_k$, namely a typical algebra over k. All group schemes are affine in this paper. We view an affine k-group scheme as a representable functor from $\mathsf{Alg}_k$ to $\mathsf{Gr}$, and we call it algebraic if the representative k-algebra is finitely generated over k. An algebraic k-group sheme is said to be <u>unipotent</u> if it admits a k-monomorphism to the k-group of all upper-triangular unipotent matrices of some fixed size (see DG-IV, §2, No. 2). With the exception of §6, all our schemes are affine over the base field and are likewise regarded as representable functors to [E].

If G is an affine k-group scheme, O(G) denotes the affine algebra representing G. O(G) is considered a Hopf algebra in the usual fashion. The underlying k-scheme of G is denoted by $\overline{G}$; thus, $\overline{G}$ = Spec O(G). <u>We refer to a k-group scheme smooth over k curtly as a k-group.</u> Therefore, our "algebraic k-group" may be regarded as a classical linear algebraic group defined over k.

For a k-scheme X and a k-algebra B, we write either $X_B$ or $X \otimes B$ to denote the B-scheme $X \times_{\text{Spec } k} (\text{Spec } B)$. Similarly for a k-group scheme G and k-algebra B.

We shall always denote by $G_a$ the one-dimensional vector group $R \mapsto G_a(R)$ := the additive group of the ring R, and by $G_m$ the one-dimensional split torus $R \mapsto G_m(R)$ := the multiplicative group of the units in R. <u>The Frobenius k-homomorphism</u> $G_a \to G_a$ given by $x \in G_a(R) \mapsto x^p \in G_a(R)$ is denoted by F. The kernel of F is denoted by $\alpha_p$, while that of F-1 is denoted by $(\mathbb{Z}/p\mathbb{Z})$.

Just to be on the safe side, let us explain some notations whose meanings should be obvious according to the current conventions and from their context: For a commutative ring A, $A^{\times}$ denotes the multiplicative group of invertible elements of A; $\mathbb{M}_m(k)$ denotes the set of all $m \times m$ matrices with entries in the field k; for a k-integral scheme X, k(X) stands for the function field of X; for any divisor D on such a scheme X, $\mathcal{O}(D) := \{f \in k(X) : (f) + D \geqslant 0\}$ ; for an arbitrary scheme Y, we write $\mathcal{O}_Y$ for its structure sheaf; when a derivation D is given on a ring A', $A'^D$ means the subring of D-constants, viz., $A'^D = \{x \in A' : D(x) = 0\}$ ; lastly, if G is a k-group functor and A is a k-G-module, the submodule of G-invariants is denoted by $A^G$ - for details on this last, consult DG-II, §§ 1-2 or SGAD-Exp. I.

We write card S for the cardinality of the set S. We have used the symbol := to mean "by definition equal to".

Finally, a list of special objects denoted by skeleton letters will follow:

$\mathbb{A}^n$ := Spec $k[T_1,\ldots,T_n]$, the affine n-space over $k$;

$\mathbb{P}^n$ := Proj $k[T_0, T_1,\ldots, T_n]$, the projective n-space

$\mathbb{Z}$ := $\{\ldots, -1, 0, 1, 2, \ldots\}$, the integers;

$\mathbb{Z}^+$ := $\{1, 2, \ldots\}$, the positive integers;

$\mathbb{N}$ := $\{0, 1, 2, \ldots\}$, the natural numbers

In the rest of this section, $k$ is assumed to have a positive characteristic $p$.

1.1. $k[F]$-modules. The endomorphism ring $\mathrm{End}_{k\text{-gr}}(G_a)$ is known to be identified as a noncommutative polynomial k-algebra with one indeterminate $F$ subject to the relation

$$F\lambda = \lambda^p F \qquad \text{for all} \quad \lambda \in k.$$

The ring $k[F]$ has a right division algorithm. To wit, given $\alpha$, $\beta \in k[F]$ with $\deg \alpha \leq \deg \beta$, there is a unique pair $(\phi, \psi)$ of elements in $k[F]$ subject to

$$\beta = \phi\alpha + \psi \quad \text{and} \quad \deg \psi < \deg \alpha ,$$

where $\deg \xi$, the degree of any $\xi \in k[F]$, is by definition the largest exponent to $F$ of all nonzero terms of $\xi$. The proof is easy by induction on $\deg \alpha$. Therefore:

1.1.1. LEMMA. Every left ideal of $k[F]$ is principal. Every left submodule of a free left $k[F]$-module is free.

If $k$ is perfect, $k[F]$ allows a left division algorithm, too. From this and from Goldie's theory on quotient rings follows the next lemma:

1.1.2. LEMMA. If $k$ is perfect, every finitely generated, torsion-free left $k[F]$-module is free. (See DG-IV, §3, 6.10 for a proof.)

1.2. Let a k-algebra $A$ be given. Turn $A$ into a left $k[F]$-module by defining $Fa := a^p$ for all $a \in A$. Let us agree that in the context of $k[F]$-module theory every k-algebra shall be considered also a left $k[F]$-module in this way. When we say that a left $k[F]$-module $M$ is a k-algebra it must be tacitly understood that $Fm = m^p$ = (the p-th power with respect to the ring multiplication of $M$) holds for every $m \in M$.

1.3. Frobenius homomorphism. Let $M$ be a left k-module, and let $A$ be a left k-algebra by $\phi : k \to A$ and a right k-algebra by $\psi : k \to A$. We denote the tensor product of the right k-algebra $A$ and the left k-module $M$ over $k$ by

$$(A,\psi) \otimes M$$

which we consider to be a left k-module via $\phi$. Thus,

$$a\psi(\lambda) \otimes m := a \otimes \lambda m, \quad \lambda(a \otimes m) := \phi(\lambda)a \otimes m$$

for all $a \in A$, $m \in M$ and $\lambda \in k$.

Typically, we take the case when $A = k$, $\phi = id_k$ and $\psi = f^n: x \mapsto x^{p^n}$ for all $x \in k$ and for a fixed $n \in \mathbb{N}$.

1.3.1. DEFINITION. For each k-module $M$, $M^{(p^n)} := (k, f^n) \otimes M$ (sometimes denoted also by $k \otimes_{p^n} M$ and its members by $\Sigma \lambda_i \otimes_{p^n} m_i$).

1.3.2. DEFINITION. For each left $k[F]$-module $M$, $F_M :=$ the mapping $M^{(p)} \to M$ defined by $F_M(\alpha \otimes_p m) := \alpha F m$ for all $\alpha \in k$, $m \in M$. $F_M$ is called <u>the Frobenius homomorphism</u> of $M$.

1.3.3. Note that $F_M$ is a k-linear mapping which is also a k-algebra homomorphism in case $M$ is a k-algebra. Furthermore, if we turn $M^{(p)}$ into a left $k[F]$-module through the definition $F(\alpha \otimes_p m) := \alpha^p \otimes_p Fm$, then $F_M$ is also a $k[F]$-module homomorphism.

1.3.4. LEMMA. <u>Let</u> $M$ <u>be a left</u> $k[F]$-<u>module. Let</u> $k' := k^{1/p}$ <u>and</u> $f : k' \to k$ <u>the homomorphism given by</u> $f(\lambda') := \lambda'^p$. <u>Consider</u> $M' := k' \otimes M$ <u>a left</u> $k[F]$-<u>module</u> via $F(\lambda' \otimes m) := \lambda'^p \otimes Fm$. <u>Then, there is a canonical isomorphism of</u> $k[F]$-<u>modules</u>

$$(k, f) \otimes_{k'} M' \simeq M^{(p)}, \qquad \alpha \otimes (\lambda' \otimes m) \longmapsto \alpha \lambda'^p \otimes m$$

*which is also a k-algebra isomorphism in case* M *is a k-algebra.*

We omit the proof of the lemma as it is a routine. Note only that $(k \hookrightarrow k' \xrightarrow{f} k)$ = (the p-th power homomorphism $k \to k$) and use the transitivity of tensor products.

1.3.5. COROLLARY. *The notations and assumptions being the same as in* 1.3.4, *we have:*

(i) $M'$ *is* $k'[F]$*-free if and only if* $M^{(p)}$ *is* $k[F]$*-free;*

(ii) *In case* M *is also a k-algebra,* $M'$ *is reduced if and only if* $M^{(p)}$ *is reduced.*

The corollary is immediate from 1.3.4 if one merely notes that $f: k' \to k$ is an isomorphism.

1.4. *Frobenius and Verschiebung homomorphisms of commutative group schemes.* Let X=Spec A be an affine k-scheme, and denote by $X^{(p)}$ the affine k-scheme Spec $A^{(p)}$. The k-algebra homomorphism $F_A: A^{(p)} \to A$ (see 1.3.2) gives a k-morphism $X \to X^{(p)}$ which we denote by $F_X$. It is called the *Frobenius k-morphism of the k-scheme* X. In case X = G = a k-group scheme, $F_G : G \to G^{(p)}$ is a k-homomorphism, too. For more details, consult DG-II, §7.

1.4.1. LEMMA. *Let* G *be an affine k-group scheme. Then,* G *is k-smooth if and only if the Frobenius k-morphism*

$F_G: G \to G^{(p)}$ is faithfully flat.

Proof (sketch). Let $\overline{G}$ = Spec A. Notice on one hand that $G$ is k-smooth if and only if $G \otimes k^{1/p}$ is reduced, and on the other that $F_G$ is faithfully flat if and only if $F_A: A^{(p)} \to A$ is an inclusion, a special feature of homomorphism of group schemes over a field. Then, the rest is taken care of by 1.3.5-(ii).

1.4.2. Let $X$ be an affine k-scheme and $\Sigma^p X$ the symmetric product of $p$ copies of $X$ obtained as the quotient of the symmetric group operation on $X^p = X \times \ldots \times X$ (p factors). There is known to be a natural k-closed immersion $X^{(p)} \hookrightarrow \Sigma^p X$ (see DG-IV, §3, 4.2). Let $G$ be a commutative k-group scheme and denote by $\pi: G^p \to G$ the k-homomorphism $p \cdot id_G$ : $g = (g_1, \ldots, g_p) \in G^p(R) \mapsto g_1 \ldots g_p \in G(R)$. Then, for every k-morphism $\phi : X \to G$ there is a unique k-morphism $\Sigma^p X \to G$ making the diagram

$$\begin{array}{ccccc} X & \xrightarrow{\Delta} & X^p & \xrightarrow{\phi^p} & G^p \\ {\scriptstyle F_X}\downarrow & & \downarrow & & \downarrow{\scriptstyle \pi} \\ X^{(p)} & \hookrightarrow & \Sigma^p X & \longrightarrow & G \end{array}$$

commutative, where $\Delta$ is the diagonal morphism (see DG-IV, §3, 4.3). The composite $X^{(p)} \to \Sigma^p X \to G$ is denoted by $\phi^V$ and is called the Verschiebung (or the shift) of $\phi$. In case

$X$ too is a commutative group scheme and $\phi$ is a homomorphism, it is easily seen that $\phi^V$ is a k-homomorphism. In the further special case of $X = G$, $\phi = id_G$, we shall write $V_G$ in place of $(id_G)^V$ and call it the Verschiebung of $G$. From these follows

1.4.3. LEMMA. For a commutative k-group scheme $G$ and the k-homomorphisms $F_G : G \to G^{(p)}$, $V_G : G^{(p)} \to G$, the identities

$$V_G F_G = p \cdot id_G \quad \text{and} \quad F_G V_G = p \cdot id_{G^{(p)}}$$

hold. (Cf. DG-IV, §3, 4.6.)

1.5. Commutative k-group schemes with null Verschiebung. Let $\boxed{V}_k$ denote the category of all commutative affine k-group schemes with null Verschiebung, and let $\boxed{Mod}_{k[F]}$ be the category of all left k[F]-modules. For each object $G \in \boxed{V}_k$, define

$$\underline{\underline{M}}(G) := Hom_{k\text{-}gr}(G, G_a)$$

which is a left $End_{k\text{-}gr}(G_a)$-module and hence a left k[F]-module in a natural fashion. $G \longmapsto \underline{\underline{M}}(G)$ gives a contravariant functor $\boxed{V}_k \to \boxed{Mod}_{k[F]}$ in an obvious manner. Next let $M$ be an arbitrary left k[F]-module, and consider it as a p-Lie algebra by defining $[m, m'] := 0$ for all $m, m' \in M$ and $m^{[p]} := Fm$. Define

$U(M)$ : = the universal enveloping k-algebra of $M$ as a p-Lie algebra,

which is further made into a Hopf algebra with antipode by introducing a comultiplication $U(M) \to U(M) \otimes U(M)$ arising from $m \longmapsto m \otimes 1 + 1 \otimes m$ ($m \in M \subseteq U(M)$, PBW Theorem !).
Let

$\underline{\underline{D}}(M)$ : = the affine k-group scheme corresponding to the Hopf algebra $U(M)$,

which gives a contravariant functor $\boxed{\text{Mod}}_{k[F]} \to \boxed{\text{V}}_k$.

1.5.1. THEOREM. <u>There is a natural isomorpnism</u>

$$\text{Hom}_{k\text{-gr}}(G, \underline{\underline{D}}(M)) \simeq \text{Hom}_{k[F]\text{-mod}}(M, \underline{\underline{M}}(G))$$

<u>functorial with respect to the variables</u> $G \in \boxed{\text{V}}_k$, $M \in \boxed{\text{Mod}}_{k[F]}$. <u>The adjoint pair</u> $(\underline{\underline{D}}, \underline{\underline{M}})$ <u>of functors gives an anti-equivalence of the categories</u> $\boxed{\text{V}}_k$ <u>and</u> $\boxed{\text{Mod}}_{k[F]}$. <u>Furthermore, under the anti-equivalence, the objects in</u> $\boxed{\text{V}}_k$ <u>which are algebraic correspond precisely to the objects in</u> $\boxed{\text{Mod}}_{k[F]}$ <u>which are finitely generated</u> $k[F]$-<u>modules</u>. (For the proof, consult DG-IV, §3, 6.7.)

1.6. A commutative k-group scheme $G$ is said to be <u>of exponent</u> $p$ if $p\cdot id_G = 0$, <u>i.e.</u>, if $x+\ldots+x$ ($p$ summands) equals $0_G$ for all $x \in G(R)$.

1.6.1. LEMMA. A k-smooth commutative k-group scheme $G$ is of exponent $p$ if and only if $V_G=0$.

Proof. By 1.4.3, $V_G = 0$ implies $p\cdot id_G = 0$. Conversely, if $p\cdot id_G = 0$ and $G$ k-smooth, then $V_G = 0$ because then $F_G$ is an epimorphism by 1.4.1.

1.6.2. Remark. Examples like $\mu_p := \operatorname{Ker} F_{G_m}$ show that the k-smoothness assumption cannot be dropped in 1.6.1. Even among the unipotent groups, $G=\alpha_p \times_\gamma G_a$ with $\gamma: \alpha_p \times \alpha_p \to G_a$ by $\gamma(\tau, \tau') := W(\tau,\tau')$, $W(X, Y) := p^{-1}[(X+Y)^p - X^p - Y^p]$ provides us an example of $G$ of exponent $p$ for which $V_G \neq 0$. (For the notations here, see Appendix, A.4.)

1.7. Let $X_0$ be an algebro-geometric object defined over k, such as a k-scheme, k-group scheme or a k-algebra. An object $X$, of the same type as $X_0$, defined over k is said to be a k-form of $X_0$ if for an algebraic extension field $k'$ of $k$ we have a $k'$-isomorphism $X \otimes k' \simeq X_0 \otimes k'$. The k-form $X$ is said to be trivial if already k-isomorphic to $X_0$. If an extension $K$ is specified, we shall mean by a $(K/k)$-form of $X_0$ any object $X$ defined over k such that $X \otimes K \simeq X_0 \otimes K$.

1.7.1. LEMMA.* A connected commutative affine algebraic k-group of exponent $p$ is a k-form of a vector group $(G_a)^m$.

*Compare Prop. 1 and Prop. 2, p.688, in "Extension of vector groups by Abelian varieties," Amer. J. Math. 80 (1958), 685-714, by M. Rosenlicht.

<u>Proof</u>. Let $G$ be a group as described. By 1.6.1, $V_G = 0$, so that $G$ is k-isomorphic to $\underline{\underline{D}}(M)$ for a suitable finitely generated $k[F]$-module $M$ by virtue of 1.5.1. One can take $\mathrm{Hom}_{k\text{-}gr}(G, G_a)$ for $M$, from which one sees that $M$ is torsion-free because $G$ is connected k-smooth. By extending the ground field to the perfect closure $k'$ of $k$ and by applying 1.1.2 one finds that $k' \otimes M$ is $k'[F]$-free and hence $G \otimes k'$ is $k'$-isomorphic to a vector group.

## 2. Forms of vector groups; groups of Russell type

*In this section (§2), the ground field* $k$ *has positive characteristic* $p$.

2.1. In this section we shall describe all left $k[F]$-modules $M$ such that $\bar{k}\otimes M$ is $\bar{k}[F]$-free of finite rank, where as always $\bar{k}$ denotes the algebraic closure of $k$. Let $M$ be such a $k[F]$-module.

2.1.1. LEMMA. *If* $k$ *is perfect,* $M$ *is free.*

*Proof.* Since $M$ is finitely generated, there is a finite field extension $K|k$ such that $K\otimes M$ is $K[F]$-free. Since any submodule of a free $k[F]$-module is free ($k[F]$ is a left PID), it is enough to show that $K[F]$ is $k[F]$-free. Let $\{e_i\}$ be a $k$-basis of $K$. Since $K|k$ is separable, $\{e_i^{p^n}\}$ is also a $k$-basis of $K$ for any integer $n \geq 0$. This means that $\{e_i\}$ forms a $k[F]$-basis of $K[F]$. (The lemma can be proven also by 1.1.2)

2.1.2. COROLLARY. *There is a finite purely inseparable extension* $K|k$ *such that* $K\otimes M$ *is* $K[F]$-*free*.

2.2. Let $M^{(p^n)} = (k,f^n)\otimes M$, where $f^n\colon k \to k$, $\lambda \mapsto \lambda^{p^n}$. The $k[F]$-linear map $M^{(p^n)} \to M$, $\lambda\otimes x \mapsto \lambda F^n x$ is injective. (To see this apply the functor $\bar{k}\otimes -$.) We shall denote its image by $M^{[n]}$. It follows from 2.1.2 that $M^{[n]}$ is $k[F]$-free for sufficiently large $n \geq 0$. In the following let $m$ be the rank of $\bar{k} \otimes M$ over $\bar{k}[F]$.

2.2.1. LEMMA. $M/M^{[n]}$ <u>is a free</u> $k[F]/F^n$<u>-module of rank</u> $m$.

<u>Proof</u>. $M/M^{[n]}$ is clearly a left $k[F]/F^n$-module and $k[F]/F^n$ is local Artinian. The dimension of $M/M^{[1]}$ over $k = k[F]/F$ is $m$, since it is left invariant by field extensions. Hence by Nakayama's Lemma, $M/M^{[n]}$ is generated by $m$ elements as a left $k[F]/F^n$-module. By counting the dimension of $M/M^{[n]}$ over $k$, which is also left invariant by field extensions, one can easily conclude that $M/M^{[n]}$ is a free $k[F]/F^n$-module of rank $m$.

2.3. Suppose that $M$ is not free. Let $n \geq 1$ be the smallest such that $M^{[n]}$ is free. Let $\{x_1,\ldots,x_m\}$ be a $k[F]$-basis of $M^{[n]}$. Let $\{\bar{y}_1,\ldots,\bar{y}_m\}$ be a $k[F]/F^n$-basis of $M/M^{[n]}$, where $y_i \in M$. Then $M$ is generated by $x_i$'s and $y_i$'s and should be determined by a set of $m$ equations of the form

$$F^n y_i = \sum_j \alpha_{ij} x_j \quad (\text{with } \alpha_{ij} \in k[F]), \quad i = 1,\ldots,m.$$

Let $A$ be the matrix $(\alpha_{ij})$ with entries in $k[F]$ and write

$A = A_0 + A_1F + \dots + A_rF^r$, $A_i \in \textcircled{M}_m(k)$. Then we can write the above relation as

$$F^nY = AX$$

where $Y = (y_1,\dots,y_m)^t$ and $X = (x_1,\dots,x_m)^t$ are column vectors.

2.3.1. LEMMA. $A_0$ <u>is invertible. One of</u> $A_1,\dots,A_r$ <u>is not contained in</u> $\textcircled{M}_m(k^p)$.

<u>Proof</u>. Since $M^{[n]}$ is generated by $F^nx_i$ and $F^ny_i$, $i = 1,\dots,m$, there are $B$ and $C \in \textcircled{M}_m(k[F])$ such that

$$X = BF^nX + CF^nY = (BF^n + CA)X.$$

Since $x_i$'s are free, it follows that $1 = BF^n + CA$. This means that $1 = C_0A_0 \in \textcircled{M}_m(k)$. Thus $A_0$ is invertible. If $A_1,\dots,A_r$ are all contained in $\textcircled{M}_m(k^p)$, it follows from 2.4.1 below (or from direct calculations) that $M^{[n-1]}$ is already a free $k[F]$-module, a contradiction.

2.4. For an integer $n \geq 0$ and an $(m \times m)$-matrix $A$ with entries in $k[F]$ let $M(n,A)$ be the left $k[F]$-module on a set of $2m$ generators $x_1,\dots,x_m$, $y_1,\dots,y_m$ defined by the following set of $m$ relations

$$F^n\begin{pmatrix} y_1 \\ \vdots \\ y_m \end{pmatrix} = A\begin{pmatrix} x_1 \\ \vdots \\ x_m \end{pmatrix}$$

for which, as in 2.3, our shorthand will be : $F^nY = AX$. In the following let us write $A = A_0 + A_1F + \ldots + A_rF^r$ with $A_i \in \mathbb{M}_m(k)$. For a matrix $C = (c_{ij})$ in $\mathbb{M}_m(k)$ put $C^{(\nu)} = (c_{ij}{}^{p^\nu})$ and let $A^{(\nu)} = A_0^{(\nu)} + A_1^{(\nu)}F + \ldots + A_r^{(\nu)}F^r$ for any $\nu \in \mathbb{Z}$.

2.4.1. PROPOSITION. (i) $M(0,A)$ <u>is a free module of rank</u> $m$.

(ii) <u>For any</u> $X \in \mathbb{GL}_m(k[F])$, $M(n,A) \simeq M(n,AX)$.

(iii) $M(n,A^{(1)}) \simeq M(n,A)^{(p)}$.

<u>In the following suppose that</u> $A_0$ <u>is invertible (that is,</u> $A_0 \in \mathbb{GL}_m(k)$).

(iv) $M(n,A)$ <u>is torsion-free.</u>

(v) $M(n,A^{(1)}) \simeq M(n-1,A)$ <u>if</u> $n > 0$.

(vi) If $A_1,\ldots,A_r \in \mathbb{M}_m(k^p)$ <u>and</u> $n > 0$, <u>then</u> $M(n,A) \simeq M(n-1,B)$ <u>for some</u> $B = B_0 + B_1F + \ldots \in \mathbb{M}_m(k[F])$ <u>with</u> $B_0 \in \mathbb{GL}_m(k)$.

(vii) $M(n,A)^{(p^n)}$ <u>is a free</u> $k[F]$-<u>module of rank</u> $m$, <u>or equivalently</u> $k^{p^{-n}} \otimes M(n,A)$ <u>is a free</u> $k^{p^{-n}}[F]$-<u>module of rank</u> $m$.

<u>Proof</u>. (i) and (ii) are clear.

(iii) Let $x_1,\ldots,x_m$, $y_1,\ldots,y_m$ be the canonical generators of $M(n,A)$. Since $F^nY = AX$, it follows that $F^n(1\otimes Y) = A^{(1)}(1\otimes X)$ in $M(n,A)^{(p)}$ where $1\otimes Y = (1\otimes y_1, \ldots, 1\otimes y_m)^t$, $1\otimes X = (1\otimes x_1,\ldots,1\otimes x_m)^t$. (Notice that $\lambda^p\otimes x = 1\otimes\lambda x$ in $M(n,A)^{(p)}$, $\lambda\in k$, $x\in M(n,A)$.) Hence we have $M(n,A)^{(p)} \simeq M(n,A^{(1)})$.

(iv) Let $x_i$, $y_i$, $1 \leq i \leq m$, be the canonical generators of $M(n,A)$. Define a filtration $\{M_\ell\}_{\ell\geq 0}$ on $M(n,A)$ as follows:

$$M_\ell = \sum_i k[F]F^\ell y_i + \sum_i k[F]x_i, \text{ if } \ell < n$$

$$M_\ell = \sum_i k[F]F^{\ell-n}x_i, \text{ if } \ell \geq n.$$

Then we have $FM_\ell \subseteq M_{\ell+1} \subseteq M_\ell$ and $\bigcap M_\ell = 0$. The induced map $\overline{F}$: $M_\ell/M_{\ell+1} \to M_{\ell+1}/M_{\ell+2}$ is injective for all $\ell \geq 0$. (For $\ell = n-1$, use the assumption that $A_0$ is invertible.) Let $\gamma = c_dF^d + c_{d+1}F^{d+1} + \ldots$ be a non-zero element in $k[F]$. Suppose that $c_d \neq 0$. Then the induced map

$$c\colon M_\ell/M_{\ell+1} \to M_{\ell+d}/M_{\ell+d+1}$$

which is equal to

$$c_dF^d\colon M_\ell/M_{\ell+1} \to M_{\ell+d}/M_{\ell+d+1}$$

is injective for all $\ell \geq 0$. This means that the map

$\gamma : M(n,A) \to M(n,A)$ is injective, since $\bigcap M_\ell = 0$. Hence $M(n,A)$ is torsion-free.

(v) Let $x_i$, $y_i$, $1 \leq i \leq m$, be the canonical generators of $M(n,A^{(1)})$. Thus we have $F^n Y = A^{(1)}X$. Define $m$ elements $z_1,\ldots,z_m$ in $M(n,A^{(1)})$ by

$$A_0 Z = F^{n-1}Y - (A_1 + \ldots + A_r F^{r-1})X$$

where $Z = (z_1,\ldots,z_m)^t$, a column vector. Then we have

$$A_0^{(1)} FZ = F^n Y - (A_1^{(1)} F + \ldots + A_r^{(1)} F^r)X = A_0^{(1)} X.$$

Hence, $FZ = X$ and $F^n Y = A^{(1)} X = FAZ$. Since $M(n,A^{(1)})$ is torsion-free, $F^{n-1}Y = AZ$; as $\{z_1,\ldots,z_m\}$ are free over $k[F]$, this proves that

$$M(n,A^{(1)}) \simeq M(n-1,A).$$

(vi) Notice that $AA_0^{-1} = B^{(1)}$ for some $B = B_0 + B_1 F + \ldots \in M_m(k[F])$ with $B_0 \in GL_m(k)$. Hence

$$M(n,A) \simeq M(n,AA_0^{-1}) \simeq M(n-1,B).$$

(vii) follows from (iii) and (v) directly.

2.4.2. Let $n, n'$ be integers $\geqq 0$ and $A, B$ be matrices with entries in $k[F]$ of the same size $m \times m$. Write $A = A_0 + A_1F + \ldots + A_rF^r$ and $B = B_0 + B_1F + \ldots + B_sF^s$ with $A_i, B_j \in M_m(k)$. Suppose $A_0, B_0 \in GL_m(k)$. Then the $k[F]$-modules $M(n,A)$ and $M(n',B)$ are $k$-forms of $k[F]^m$. We shall give a necessary and sufficient condition in order that $M(n,A)$ and $M(n',B)$ be isomorphic to each other.

First, in view of Proposition 2.4.1 (v), we have

$$M(n,A) \simeq M(n+n',A^{(n')}) \quad \text{and}$$

$$M(n',B) \simeq M(n+n',B^{(n)}).$$

Therfore we may and shall assume that $n = n'$

2.4.3. THEOREM. Let $n$ be an integer $\geqq 0$ and $A, B \in M_m(k[F])$ be such that $A_0, B_0 \in GL_m(k)$. Then the $k[F]$-modules $M(n,A)$ and $M(n,B)$ are isomorphic to each other if and only if there exist matrices $X \in GL_m(k[F])$ and $Y, Z \in M_m(k[F])$ such that

$$B = Y^{(n)}AX + F^nZ.$$

Here the matrix $Y$ can be so chosen of the form

$$Y = Y_0 + Y_1F + \ldots + Y_{n-1}F^{n-1}$$

with $Y_0 \in GL_m(k)$ and $Y_i \in M_m(k)$.

Proof. Suppose that $M(n,A) = M = M(n,B)$. Let $\{x_1,\ldots,x_m,y_1,\ldots,y_m\}$ and $\{u_1,\ldots,u_m,v_1,\ldots,v_m\}$ denote the canonical systems of generators of $M(n,A)$ and $M(n,B)$ respectively.

Thus we have

$$F^n y = Ax \quad \text{and} \quad F^n v = Bu$$

where $x = (x_1,\ldots,x_m)^t$, $y = (y_1,\ldots,y_m)^t$, etc. The submodule $M^{[n]}$ of $M$, which is isomorphic with $M^{(p^n)}$, is a free $k[F]$-module of rank $m$ by Proposition 2.4.1 (vii). It is easy to see that $\{x_1,\ldots,x_m\}$ and $\{u_1,\ldots,u_m\}$ form two free $k[F]$-bases of $M^{[n]}$ (cf. the proof of Proposition 2.4.1 (v)). On the other hand, $M/M^{[n]}$ is a free $k[F]/F^n$-module of rank $m$ by Lemma 2.2.1. Since $M/M^{[n]}$ is generated by $\{\overline{y_1},\ldots,\overline{y_m}\}$, by counting the dimension over $k$ one concludes that $\{\overline{y_1},\ldots,\overline{y_m}\}$ and $\{\overline{v_1},\ldots,\overline{v_m}\}$ form two free $k[F]/F^n$-bases of $M/M^{[n]}$. Therefore there exist matrices $X \in GL_m(k[F])$ and $Y \in M_m(k[F])$ such that

$$x = Xu \quad \text{and} \quad v \equiv Yy \mod. M^{[n]}$$

and hence we have

$$v = Yy + Zu$$

for some $Z \in M_m(k[F])$. Here we can take the matrix $Y$ to be of the form

$$Y = Y_0 + Y_1 F + \ldots + Y_{n-1} F^{n-1}$$

with $Y_0 \in GL_m(k)$ and $Y_i \in M_m(k)$. Now we have

$$Bu = F^n v = Y^{(n)} F^n y + F^n Zu = (Y^{(n)} AX + F^n Z)u$$

and hence

$$B = Y^{(n)}AX + F^nZ.$$

Conversely suppose that $B = Y^{(n)}AX + F^nZ$ for some matrices $X \in GL_m(k[F])$ and $Y, Z \in M_m(k[F])$ and let $\{x_1, \ldots, x_m, y_1, \ldots, y_m\}$ denote the canonical generators of $M(n,A)$. In order to show $M(n,A) \approx M(n,B)$, we can assume that $n > 0$. Since then

$$B_0 = Y_0^{(n)}A_0X_0,$$

it follows that $Y_0 \in GL_m(k)$ and hence that $\bar{Y} \in GL_m(k[F]/F^n)$. If we put $u = X^{-1}x$ and $v = Yy + Zu$, then the $k[F]$-module $M(n,A)$ is generated by $\{u_1, \ldots, u_m, v_1, \ldots, v_m\}$ with the defining relation $F^nv = Bu$. This proves that $M(n,A) \approx M(n,B)$.

If $m = 1$, then $GL_1(k[F]) = k^{\times}$. Hence our result coincides with [11;Prop.2.3] in this case.

2.4.4. A pair $(n,A)$ with $n \in \mathbb{N}$ and $A \in M_m(k[F])$ will be called admissible if either (i) $n = 0$ or (ii) $A_0 \in GL_m(k)$ and $AX \notin M_m(k^p[F])$ whenever $X \in GL_m(k[F])$. Let $M$ be a $k$-form of $k[F]^m$. By the height of $M$, we mean the smallest integer $n \geqq 0$ such that $M^{(p^n)} \approx k[F]^m$.

COROLLARY. (i) *Every k-form of* $k[F]^m$ *is isomorphic with* $M(n,A)$ *for some admissible pair* $(n,A)$.

(ii) *If* $(n,A)$ *is an admissible pair, then* $n$ *is the height of* $M(n,A)$. *Hence if* $(n',A')$ *is another admissible pair, then* $M(n,A) \approx M(n',A')$ *implies* $n = n'$.

2.5. Conclusion. For an integer $n \geqq 0$ and a matrix $A = A_0 + A_1F + \ldots + A_rF^r \in M_m(k[F])$ with $A_0 \in GL_m(k)$ and

$A_i \in \mathbb{M}_m(k)$, the left $k[F]$-module $M(n,A)$ on a set of $2m$ generators $\{x_1,\ldots,x_m,y_1,\ldots,y_m\}$ defined by

$$F^n \begin{pmatrix} y_1 \\ \vdots \\ y_m \end{pmatrix} = A \begin{pmatrix} x_1 \\ \vdots \\ x_m \end{pmatrix}$$

is a $(k^{p^{-n}}/k)$-form of $k[F]^m$. Conversely, if a $k[F]$-module $M$ is a $(\bar{k}/k)$-form of $k[F]^m$, then there is a pair $(n,A)$ as above such that $M \simeq M(n,A)$. If $M$ is not itself free, then the matrix $A$ can be so chosen that $AX \notin \mathbb{M}_m(k^p[F])$ whenever $X \in GL_m(k[F])$. Once $A$ is so chosen, the integer $n$ is uniquely determined by $M$; $n$ is the smallest integer such that $M^{(p^n)}$ is $k[F]$-free.

2.6. In light of 1.5.1, the conclusion 2.5 above admits a more algebro-geometric interpretation: If a k-group scheme $G$ is a k-form of the vector group $(G_a)^m$, then $G$ is affine, algebraic, commutative, k-smooth and of exponent $p$; hence, by 1.6.1, $G$ has null Verschiebung (viz., $V_G = 0$) and $\bar{k} \otimes \underline{\underline{M}}(G)$ is $\bar{k}[F]$-free of rank $m$. By the foregoing theory, $\underline{\underline{M}}(G) \simeq M(n,A)$ and $G \simeq \underline{\underline{D}}(M(n,A))$. Conversely, for each $k[F]$-module $M(n,A)$ as in 2.5, the k-group $\underline{\underline{D}}(M(n,A))$ is a k-form of a vector group. <u>We have thus determined all</u>

k-*forms of the vector group* $(G_a)^m$, generalizing Russell's result [11].

2.7. Consider the case $m = 1$, treated by Russell (*ibid.*). A pair $(n,\alpha)$ with $n \in \mathbb{N}$ and $\alpha = \sum a_i F^i \in k[F]$ will be called *admissible* if either (i) $n = 0$ or (ii) $a_0 \neq 0$ and $a_i \notin k^p$ for some $i > 0$. By a k-*group of Russell type* we mean a one-dimensional unipotent group in the form of $\underline{D}(M(n,\alpha))$ where $(n,\alpha)$ is admissible *and* $n > 0$. By what we saw just now, *the* k-*groups of Russell type are all and only nontrivial* k-*forms of* $G_a$, up to k-isomorphisms. Let $\underline{D}(M(n,\alpha))$ be a k-group of this type; it therefore represents a k-closed subgroup scheme of $(G_a)^2$ whose underlying scheme is given in $\mathbb{A}^2$ by the equation

$$Y^{p^n} = a_0 X + a_1 X^p + \ldots + a_r X^{p^r}$$

where $a_0 \neq 0$ and $a_i \notin k^p$ for some $1 \leq i \leq r$. The k[F]-module $M(n,\alpha)$ is the factor module of the free module $(k[F])^2$ by the single relation $F^n(0,1) = \alpha(1,0)$.

2.8. We prove at this point a few preparatory results on Russell type groups.

2.8.1. LEMMA. *Let* $(n,\alpha)$, $(\ell,\beta)$ *be admissible pairs. If* $n > \ell$, *then* $\mathrm{Hom}_{k[F]}(M(n,\alpha), M(\ell,\beta)) = 0$.

Proof. Since $M(n,\alpha)^{(p^{n-1})} \simeq M(1,\alpha)$ and $M(\ell,\beta)^{(p^{n-1})} \simeq k[F]$, we have only to show that $\mathrm{Hom}_{k[F]}(M(1,\alpha),k[F]) = 0$. Indeed let $\xi$ and $\eta \in k[F]$ be such that $\alpha\xi = F\eta$. Since one of $a_1,\ldots,a_r$ is not in $k^p$, it follows that $\xi = \eta = 0$.

2.8.2. COROLLARY. Let $(n,\alpha)$ with $n \geq 0$ and $\alpha \in k[F]$ be an admissible pair. Then $n$ is the smallest integer such that $M(n,\alpha)^{(p^n)} \simeq k[F]$.

Proof. If $M(n,\alpha)^{(p^\ell)} \simeq k[F]$ for some $\ell < n$, then we have $M(n-\ell,\alpha) \simeq k[F]$, a contradiction.

2.8.3. LEMMA. Let $(n,\alpha)$ and $(\ell,\beta)$ be two admissible pairs where $n,\ell \geq 0$ and $\alpha,\beta \in k[F]$. If $n > \ell$, then $\mathrm{Hom}_{k[F]}(M(n,\alpha), \Lambda^2 M(\ell,\beta)) = 0$.*

Proof. Applying the functor $M \mapsto M^{(p^{n-1})}$, we have only to show that $\mathrm{Hom}_{k[F]}(M(1,\alpha), \Lambda^2 k[F]) = 0$. It is easy to see $F(\Lambda^2 k[F]) \cap \alpha(\Lambda^2 k[F]) = 0$. Since $\Lambda^2 k[F]$ is torsion-free, the assertion follows.

---

* For $k[F]$-modules $M$, $N$, the $k$-modules $M \otimes N$ and $M \wedge N$ are made into $k[F]$-modules via $F(m \otimes n) := Fm \otimes Fn$, $F(m \wedge n) := Fm \wedge Fn$ for $m \in M$, $n \in N$.

## 3. Decomposition theorems for central extensions of comutative group schemes; application to the two-dimensional unipotent groups

In this section (§3), at first (in 3.1 and 3.2) k is an arbitrary field; from 3.3 until the end, k is assumed to have a positive characteristic p. All group schemes are assumed to be affine. For the generalities on central extensions, see Appendix below.

3.0. In this section, we first prove a direct sum decomposition theorem of $\mathrm{Ext}_{\mathrm{cent}}(B,A)$ for arbitrary commutative k-group schemes A, B (see 3.2). We then make the result more elaborate in case $A = \underline{\underline{D}}(M)$, $B = \underline{\underline{D}}(N)$ for some k[F]-modules M, N (see 3.4.1). Finally, we shall show how these results in conjunction with our §2 yield a classification of all two-dimensional unipotent k-groups.

3.1. Let A, B be commutative k-group schemes, and denote by $\mathcal{B}(B,A)$ the set of all biadditive k-morphisms from $B \times B$ to A. Thus, a k-morphism $f: B \times B \to A$ belongs to $\mathcal{B}(B,A)$ if and only if $f(x+y, z) = f(x,z) + f(y,z)$, $f(x,y+z) = f(x,y) + f(x,z)$ hold for all $x, y, z \in B(R)$. Further, denote by $\mathcal{B}^{\circ}(B,A)$ the set of those $f \in \mathcal{B}(B,A)$ which are antisymmetric, i.e., which satisfy $f(x,x) = 0$ for all $x \in B(R)$. (This last implies $f(x,y) =$

$-f(y,x)$, while the converse is true provided $2\cdot id_A$ is a monomorphism.) By defining $f+g$ through the rule $(f+g)(x,y):=f(x,y)+g(x,y)$ for all $x,y \in B(R)$, we turn the sets $\mathcal{B}(B,A)$ and $\mathcal{B}^\circ(B,A)$ into additive groups. Moreover, each has a natural structure of left $End_{k\text{-}gr}(A)$-module.

3.2. THEOREM <u>Let</u> A, B <u>be commutative</u> k-<u>group schemes</u>.

(i) <u>There is an exact sequence of left</u> $End_{k\text{-}gr}(A)$-<u>modules</u> (<u>arising from the commutator function</u>):

$$0 \to Ext_{com}(B,A) \to Ext_{cent}(B,A) \to \mathcal{B}^\circ(B,A). \tag{1}$$

(ii) <u>If</u> $2\cdot id_A$ <u>has a right inverse in</u> $End_{k\text{-}gr}(A)$, <u>then</u> (1) <u>is split exact to give the direct sum decomposition</u>:

$$Ext_{cent}(B,A) \simeq Ext_{com}(B,A) \oplus \mathcal{B}^\circ(B,A) \tag{2}$$

<u>as</u> $End_{k\text{-}gr}(A)$-<u>modules</u>.

<u>Proof</u>. (i) Let $0 \to A \to G \to B \to 0$ be a central extension, and consider the commutator function $[-,-]$: $G \times G \to G$ defined by $(x,y) \in G(R) \times G(R) \mapsto xyx^{-1}y^{-1} \in G(R)$. Since B is commutative, it gives a k-morphism $G \times G \to A$, which factors through the canonical k-morphism $G \times G \to B \times B$ because A is central. As a result, one obtains a k-morphism $\gamma_G$: $B \times B \to A$. One can verify without difficulty that $\gamma_G$

is antisymmetric and biadditive. Also easy is to see that $\gamma_G = \gamma_{G'}$ if $G$ and $G'$ are equivalent central extensions of $B$ by $A$. Thus, $G \longmapsto \gamma_G$ defines a mapping of $\mathrm{Ext}_{\mathrm{cent}}(B,A)$ into $\mathcal{B}^\circ(B,A)$. By remembering the make-up of the additive group structure on $\mathrm{Ext}_{\mathrm{cent}}$, one can also see that $G \longmapsto \gamma_G$ gives a homomorphism of additive groups $\mathrm{Ext}_{\mathrm{cent}}(B,A) \to \mathcal{B}^\circ(B,A)$. Finally, for any $\phi \in \mathrm{End}_{k\text{-}gr}(A)$, one computes the commutator function on $\phi_* G$ and quickly discovers that it is the commutator function on $G$ followed by $\phi$ itself.

(ii) Assume now that $2 \cdot \mathrm{id}_A$ has a right inverse $(1/2)$. Let $f \in \mathcal{B}^\circ(B,A)$ and define on the k-scheme $B \times \overline{A}$ a group structure by means of the formula

$$(b,a)(b',a') = (b+b',\ a+a'+(1/2)f(b,b'))$$

for all $a,a' \in \overline{A}(R)$ and for all $b,b' \in \overline{B}(R)$. Since every biadditive k-morphism $B \times B \to A$ satisfies the usual cocycle condition, $\overline{B} \times \overline{A}$ thus becomes a k-group scheme which we denote by $B \times_{(1/2)f} A$. We have an obvious, geometrically k-split central extension $0 \to A \to B \times_{(1/2)f} A \to B \to 0$ for which we calculate the corresponding biadditive function on $B \times B$. It turns out to be equal to $(1/2)f + (1/2)f = 2((1/2)f) = f$. Thus, $f \longmapsto$ (the extension class of $B \times_{(1/2)f} A$) gives the desired splitting of (1) and we have established (2).

Q.E.D.

3.3. Let $L$, $M$ and $N$ be left $k[F]$-modules. We view $M \otimes N$ as a left $k[F]$-module by $F(x \otimes y) = Fx \otimes Fy$, $x \in M$, $y \in N$. Let $f: L \to M \otimes N$ be a $k[F]$-linear map. Extend this to an algebra map $\bar{f}: U(L) \to U(M) \otimes U(N)$. If we view $\bar{f}$ as a morphism $\underline{\underline{D}}(M) \times \underline{\underline{D}}(N) \to \underline{\underline{D}}(L)$, then $\bar{f}$ is easily seen to be biadditive. Conversely,

3.3.1. LEMMA. Every bi-additive morphism $\underline{\underline{D}}(M) \times \underline{\underline{D}}(N) \to \underline{\underline{D}}(L)$ comes from a $k[F]$-linear map $M \otimes N \leftarrow L$.

Proof. Let $X$, $Y$ and $Z$ be commutative k-group schemes and let $A = O(X)$, $B = O(Y)$, $C = O(Z)$ be their affine Hopf algebras. Let $f: X \times Y \to Z$ be a bi-additive k-morphism. We identify $f$ with an algebra map $f: C \to A \otimes B$. The map $X \times Y \to Z \times Y$, $(x,y) \longmapsto (f(x,y),y)$, is clearly a Y-group map, or equivalently the map

$$C \otimes B \to A \otimes B, \; c \otimes b \longmapsto f(c)(1 \otimes b)$$

is a B-Hopf algebra map. Let $P(C)$ denote the primitive elements in $C$. Since the primitive elements of the B-Hopf algebra $A \otimes B$ are $P(A) \otimes B$, it follows from above that $f(P(C)) \subseteq P(A) \otimes B$. Similarly we have $f(P(C)) \subseteq A \otimes P(B)$. Hence $f(P(C)) \subseteq P(A) \otimes P(B)$. Since $P(U(M)) = M$, this means that any bi-additive morphism $\underline{\underline{D}}(M) \times \underline{\underline{D}}(N) \to \underline{\underline{D}}(L)$ comes from a $k[F]$-linear map $L \to M \otimes N$.

3.4. Let $G_1 = \underline{D}(M_1)$ and $G_2 = \underline{D}(M_2)$, where $M_1$ and $M_2$ are left $k[F]$-modules. Define $E^2M_2$ as follows: $E^2M_2$ is the anti-symmetric elements in $M_2 \otimes M_2$ if $p \neq 2$; $E^2M_2$ is the k-subspace of $M_2 \otimes M_2$ generated by $x \otimes y + y \otimes x$, $x, y \in M_2$, and $\sum_i \xi_i x_i \otimes x_i$, $\xi_i \in k$, $x_i \in M_2$ such that $\sum \xi_i F x_i = 0$, if $p = 2$. This is clearly a $k[F]$-submodule of $M_2 \otimes M_2$.

3.4.1. THEOREM. *With the notations as above, there are two exact sequences:*

$$0 \to \mathrm{Ext}_{com}(G_2,G_1) \to \mathrm{Ext}_{cent}(G_2,G_1) \xrightarrow{\pi} \mathrm{Hom}_{k[F]}(M_1,E^2M_2)$$

$$0 \to \mathrm{Ext}^1_{k[F]}(M_1,M_2) \xrightarrow{\rho} \mathrm{Ext}_{com}(G_2,G_1) \xrightarrow{\sigma} \mathrm{Hom}_{k[F]}(M_1,M_2^{(p)}).$$

(i) *If* $p \neq 2$, $\pi$ *has a natural section*.

(ii) *If* $M_1$ *is finitely generated*, $\sigma$ *is surjective*.

(iii) *If* $M_2$ *is finitely generated*, $\sigma$ *has a section which is natural with respect to* $M_1$.

(iv) *Hence if* $p \neq 2$ *and* $M_2$ *is finitely generated, then we have*

$$\mathrm{Ext}_{cent}(G_2,G_1) \simeq \mathrm{Ext}^1_{k[F]}(M_1,M_2) \oplus \mathrm{Hom}_{k[F]}(M_1,M_2^{(p)} \oplus \Lambda^2 M_2)$$

*which is natural with respect to* $M_1$.

<u>Proof</u>. <u>Construction</u> <u>of</u> $\pi$, $\rho$ <u>and</u> $\sigma$: By virtue of 3.2-(i) and 3.3.1, we can construct an exact sequence:

$$0 \to \mathrm{Ext}_{\mathrm{com}}(G_2,G_1) \to \mathrm{Ext}_{\mathrm{cent}}(G_2,G_1) \xrightarrow{\pi} \mathrm{Hom}_{k[F]}(M_1,M_2 \otimes M_2).$$

For each extension class represented by $E$, we must show that the image of $\pi(E)$: $M_1 \to M_2 \otimes M_2$ is contained in $E^2M_2$. This is clear if $p \neq 2$, since $\pi(E)$ is anti-symmetric. If $p = 2$, the composite

$$M_1 \xrightarrow{\pi(E)} M_2 \otimes M_2 \to U(M_2)$$

$$x \otimes y \longmapsto xy$$

is clearly zero. The Poincaré-Birkhoff-Witt theorem implies that the kernel of $M_2 \otimes M_2 \to U(M_2)$ given by $x \otimes y \longmapsto xy$ is exactly $E^2M_2$. Thus we get the first exact sequence.

Next suppose given a commutative extension $0 \to G_1 \to E \to G_2 \to 0$ (<u>i.e.</u>, $E$ is commutative). Since the $G_i$ are killed by the Verschiebung, the Verschiebung homomorphism of $E$ factors as

$$V_E\colon E^{(p)} \to G_2^{(p)} \xrightarrow{\sigma(E)} G_1 \to E$$

and we may view $\sigma(E)$ as a $k[F]$-linear map $M_1 \to M_2^{(p)}$. The map $\mathrm{Ext}_{\mathrm{com}}(G_2,G_1) \to \mathrm{Hom}_{k[F]}(M_1,M_2^{(p)})$ given by $E \longmapsto \sigma(E)$ also is a homomorphism whose kernel is exactly the image of the map

$$\rho\colon \mathrm{Ext}^1_{k[F]}(M_1,M_2) \to \mathrm{Ext}_{\mathrm{com}}(G_2,G_1)$$

defined by $(0 \to M_2 \to M \to M_1 \to 0) \longmapsto (0 \to G_1 \to \underline{D}(M) \to G_2 \to 0)$. Since $\rho$ is clearly injective, we have the second exact sequence. Now, in more detail:

(i) In case $p \neq 2$, we have the identifications $E^2M_2 = \Lambda^2 M_2$ and $\mathrm{Hom}_{k[F]}(M_1,E^2M_2) = \mathcal{B}^{\circ}(G_2,G_1)$. Also, in this case, $2\cdot\mathrm{id}_{G_1}$ is clearly an isomorphism in $\mathrm{End}_{k\text{-}gr}(G_1) = \mathrm{End}_{k[F]}(M_1)$ (notice $2F = F2$). Therefore, by 3.2-(ii), we have a splitting of our exact sequence.

(ii) First consider the case $M_1 = k[F]$. Since $\mathrm{Ext}^1_{k[F]}(k[F],\, M_2) = 0$, we have $\sigma\colon \mathrm{Ext}_{\mathrm{com}}(G_2,G_a) \hookrightarrow \mathrm{Hom}_{k[F]}(k[F],\, M_2^{(p)}) = M_2^{(p)}$. The naturality of $\sigma$ with respect to $M_1$ implies that this is $k[F]$-linear. Hence, in order to show that this is surjective, it suffices to see $1 \otimes u \in \mathrm{Im}(\sigma)$ for all $u \in M_2$. Now let

$$(E_0) \qquad 0 \to G_a \to G_a \times_{-W} G_a \to G_a \to 0$$

be the central extension determined by the 2-cocycle $-W = (-1/p)[(X+Y)^p - X^p - Y^p]$. It is well-known that the

Verschiebung of $G_a \times_{-W} G_a$ is $(x,a) \longmapsto (0,x)$. This means that $\sigma(E_0) = 1 \otimes 1 \in k[F]^{(p)}$. Let $u: G_2 \to G_a$ be the group homomorphism associated with $k[F] \to M_2$, $1 \longmapsto u$. Then we have clearly $\sigma(E_0 u) = 1 \otimes u$. Hence we have

$$\sigma: \mathrm{Ext}_{\mathrm{com}}(G_2, G_a) \overset{\approx}{\to} M_2^{(p)}.$$

In general every left $k[F]$-module has a free resolution of length $\leq 2$, since $k[F]$ is a left principal ideal domain [1, page 81, Remark]. If $M_1$ is finitely generated, there is a finitely generated free resolution $0 \to P \to Q \to M_1 \to 0$. Observe the following commutative diagram:

$$\begin{array}{ccccccc} & & & & & & 0 \\ & & & & & & \downarrow \\ 0 \to & \mathrm{Ext}^1_{k[F]}(M_1, M_2) & \overset{\rho}{\to} & \mathrm{Ext}_{\mathrm{com}}(G_2, G_1) & \overset{\sigma}{\to} & \mathrm{Hom}_{k[F]}(M_1, M_2^{(p)}) \\ & & & \downarrow & & \downarrow \\ & & & \mathrm{Ext}_{\mathrm{com}}(G_2, \underline{D}(Q)) & \overset{\approx}{\to} & \mathrm{Hom}_{k[F]}(Q, M_2^{(p)}) \\ & & & \downarrow & & \downarrow \\ & & & \mathrm{Ext}_{\mathrm{com}}(G_2, \underline{D}(P)) & \overset{\approx}{\to} & \mathrm{Hom}_{k[F]}(P, M_2^{(p)}). \end{array}$$

All of its rows and columns are exact (notice that $0 \to G_1 \to \underline{D}(Q) \to \underline{D}(P) \to 0$ is exact). Therefore one sees immediately that the map $\sigma$ is surjective.

(iii) Since $M_2^{(p)}$ is finitely generated, it follows from (ii) that there is a commutative extension $E: 0 \to G_2^{(p)}$

$\to E \to G_2 \to 0$ such that $\sigma(E) = 1 \in \mathrm{Hom}_{k[F]}(M_2^{(p)}, M_2^{(p)})$. Then the map

$$\mathrm{Hom}_{k[F]}(M_1, M_2^{(p)}) \to \mathrm{Ext}_{com}(G_2, G_1) \quad \text{given by} \quad f \longmapsto \underline{D}(f)E$$

is a section of $\sigma$ and natural with respect to $M_1$. (Notice that $\underline{D}(f) \in \mathrm{Hom}_{k\text{-}gr}(G_2^{(p)}, G_1)$ for $f \in \mathrm{Hom}_{k[F]}(M_1, M_2^{(p)})$.)

(iv) This is clear since $\Lambda^2 M_2 \cong E^2 M_2$ via $x \wedge y \longmapsto x \otimes y - y \otimes x$ if $p \neq 2$.

3.5. We saw in §2 that if $G_1$ is a k-form of $G_a$ then $M_1$ is generated by two elements $u$ and $v$ with one relation $F^n u = \alpha v$, $\alpha \in k[F]$. For such an $M_1$, it is easy to check

$$\mathrm{Ext}^1_{K[F]}(M_1, M_2) \simeq M_2/(F^n M_2 + \alpha M_2).$$

3.6. For a left $k[F]$-module $M$ and an integer $n \geq 0$, let ${}_nM$ denote the kernel of $F^n: M \to M$. Then $\mathrm{Hom}_{k[F]}(k[F]/F^n, M) \simeq {}_nM$. Notice that $\alpha_{p^n} = \underline{D}(k[F]/F^n)$.

3.6.1. Examples. $\mathrm{Ext}_{com}(G_2, G_a) \simeq M_2^{(p)}$. $\mathrm{Ext}_{com}(G_2, \alpha_{p^n}) \simeq M_2/F^n M_2 \oplus {}_n(M_2^{(p)})$ if $M_2$ is f. g.. $\mathrm{Ext}_{cent}(G_2, G_a) \simeq M_2^{(p)} \oplus \Lambda^2 M_2$ if $p \neq 2$. $\mathrm{Ext}_{cent}(G_2, \alpha_{p^n}) \simeq M_2/F^n M_2 \oplus {}_n(M_2^{(p)}) \oplus {}_n(\Lambda^2 M_2)$ if further $M_2$ is f. g. (cf. DG-II, §3,

4.6, III, §6,5.3, 7.6, 7.7 ).

3.7. Two-dimensional unipotent groups. It is well known (cf. §1 & DG-IV, §2) that connected 1-dimensional unipotent k-groups are k-forms of $G_a$. Hence they come from some admissible pair $(n,\alpha)$ with $n \geq 0$ and $\alpha \in k[F]$: $G \simeq \underline{D}(M(n,\alpha))$.

Let G be a connected 2-dimensional unipotent k-group. Then clearly G is of one of the following three types:

i) G is not commutative.

ii) G is commutative and $V_G$ (= the Verschiebung of G) $\neq 0$.

iii) G is commutative and $V_G = 0$.

In case i) let $G_1 = [G,G]$ be the commutator subgroup of G. In case ii) let $G_1 = V_G(G^{(p)})$ be the image of the Verschiebung map. Then $G_1$ and $G_2 = G/G_1$ are connected k-smooth 1-dimensional unipotent k-group schemes and hence k-forms of $G_a$. Hence there are two admissible pairs $(n_i,\beta_i)$ with $n_i \geq 0$ and $\beta_i \in k[F]$ such that $G_i \simeq \underline{D}(M_i)$ where $M_i = M(n_i,\beta_i)$ and we have a central extension

$$0 \to G_1 \to G \to G_2 \to 0.$$

Suppose that $p \neq 2$ for simplicity. Then we have

$$\mathrm{Ext}_{cent}(G_2,G_1) \simeq M_2/(F^{n_1}M_2+\beta_1 M_2) \oplus \mathrm{Hom}_{k[F]}(M_1, M_2^{(p)} \oplus \Lambda^2 M_2)$$

by 3.4 and 3.5. In case i) we have $\mathrm{Hom}_{k[F]}(M_1,\Lambda^2 M_2) \neq 0$. Hence $n_1 \leq n_2$ by 2.8.3. In case ii) we have $\mathrm{Hom}_{k[F]}(M_1,M_2^{(p)}) \neq 0$. Since $M_2^{(p)} \simeq M(n_2 - 1,\beta_2)$ if $n_2 > 0$, it follows that $n_1 < n_2$ or $n_1 = 0$ from 2.8.1. In either case the possibility $n_1 > n_2$ cannot happen.

In case iii) G is by 1.7.1 a k-form of $(G_a)^2$. Hence there is a pair $(n,A)$ with $n \geq 0$ and $A = A_0 + A_1 F + \cdots \in \underline{M}_2(k[F])$ where $A_0 \in \underline{GL}_2(k)$ and $A_1,\cdots \in \underline{M}_2(k)$ such that $G \simeq \underline{\underline{D}}(M(n,A))$. In rather a formal sense, the foregoing is a complete description of the groups of the envisaged type in terms of $k[F]$-modules.

## 4. Wound unipotent groups

*In this section (§4), the ground field* $k$ *has a positive characteristic* $p$.

4.1. PROPOSITION. *Let* $f : \mathbb{A}^1 \longrightarrow X = \operatorname{Spec}(k[x_1, \ldots, x_n])$ *be a nonconstant k-morphism, and decompose it as* $f = (\mathbb{A}^1 \twoheadrightarrow X_0 \hookrightarrow X)$ *where* $\mathbb{A}^1 \twoheadrightarrow X_0$ *is the epimorphism to the image* $X_0$ *of* $f$ *and* $X_0 \hookrightarrow X$ *is the closed immersion. Then*

(i) $\mathbb{A}^1 \twoheadrightarrow X_0$ *is a finite morphism and* $\mathbb{A}^1$ *is the normalization of* $X_0$ *in* $k(\mathbb{A}^1)$;

(ii) $X_0$ *is a rational curve and the only place of* $k(X_0)$ *not finite on* $X_0$ *comes from the infinite place of* $k(\mathbb{A}^1)$; *and*

(iii) $X_0$ *is k-isomorphic with* $\mathbb{A}^1$ *if and only if* $X_0$ *is k-normal.*

*Proof.* Only the 'if' part of (iii) requires a proof. Let $k(X_0) = k(v)$, rational function field in one variable $v$ over $k$. Then, $v = g(t)/h(t) \in k(\mathbb{A}^1)$, and $t \longrightarrow \infty$ gives the only place not finite on $X_0$. If $v \longrightarrow a \in k$ under that place, set $u = 1/(v-a)$; if $v \longrightarrow \infty$, set $u = v$. Then it is clear that $k[X_0] = k[u]$.

4.2. COROLLARY. *Every quotient of* $G_a$ *is k-isomorphic with* $G_a$, *and the kernel of the canonical homomorphism is given as the zeroes of a p-polynomial over* k.

*Proof*. The first assertion follows from 4.1-(iii), and the second is verified by the well-known calculation.

The corollary above is, of course, very well known.

4.3. Let G be a commutative k-group of exponent p. As noted in §1, G is then a k-closed subgroup of a vector group. It is well known (cf. [15; III, 3.3, p.120 ff], [9; PROP., p.102]) that the defining equations of G consist of p-polynomials in several variables. (In case k is infinite, which is really the case that counts, one may take $\bar{G}$ to be of codimension one in the ambient vector group—see Tits [15; III, 3.3.1]. We do not use this result, though.)

4.3.1. THEOREM. *Let* G *be a unipotent algebraic* k-*group*. *Then, the following are equivalent*:

(i) *Every* k-*homomorphism* $G_a \longrightarrow G$ *is constant*;

(ii) G *contains no* k-*closed subgroup* k-*isomorphic to* $G_a$;

(iii) *Every* k-*morphism* $\mathbb{A}^1 \longrightarrow \bar{G}$ *is constant*;

(iv) $\bar{G}$ *contains no* k-*closed subscheme* k-*isomorphic to* $\mathbb{A}^1$.

The theorem is essentially due to Tits [15; IV, 4, p.151 ff]. We offer a simpler proof which, however, does not yield the decomposition theorem in its course as did Tits' original proof (*ibid*., IV, 4.2).

Observe first that (iii)$\Longrightarrow$(iv)$\Longrightarrow$(ii) is trivially true and (ii) $\Longrightarrow$(i) is immediate from 4.2. Therefore, what remains to be shown is the implication (i)$\Longrightarrow$(iii). We shall conduct its proof by proving two lemmas:

4.4 LEMMA. Let $k'/k$ be a Galois extension with Galois group $\Gamma$, and let $G$ be a commutative $k$-group scheme. Then, the set $H = \mathrm{Hom}_{k'\text{-gr}}((G_a)_{k'}, G_{k'})$ has a natural right $k'$-module structure and also admits a natural action of $\Gamma$ on itself in such a manner that

$$(f\alpha)^{\sigma} = f^{\sigma}\alpha^{\sigma} \quad \text{for all } \alpha \in k',\ \sigma \in \Gamma, \text{ and } f \in H.$$

Furthermore, the $\Gamma$-invariant elements $H^{\Gamma}$ of $H$ coincides with $\mathrm{Hom}_{k\text{-gr}}(G_a, G)$ which is a $k$-form of the $k'$-module $H$.

The first half of the lemma hardly requires proof, while the second half is the standard Galois descent (see, e.g., [15; I, 5.3.2, p. 42, and Errata, p.1]).

4.5. LEMMA. Let $G$ be a commutative $k$-group of exponint $p$, and let there exist $\psi : \mathbb{A}^1 \longrightarrow G$, a nonconstant $k$-morphism. Then, there exists a nonconstant $k$-homomorphism $\phi : G_a \longrightarrow G$.

Proof. By what we remarked in 4.3, we may consider $G$ as a $k$-closed subgroup of $G_a \times \ldots \times G_a$ ($n$ factors) defined by the equations $\Phi_\nu = 0$ ($\nu \in N$ = an index set), where each $\Phi_\nu \in k[X_1, \ldots, X_n]$ is a $p$-polynomial; thus, for each $\nu$, one can write

$$\Phi_\nu = \Phi_{\nu 1} + \ldots + \Phi_{\nu n}, \quad \Phi_{\nu i} = \sum_{j \geq 0} a_{ij}^{(\nu)} X_i^{p^j}.$$

By assumption, there exist $f_1, \ldots, f_n \in k[T]$ of which at least one is nonconstant such that

$$\Phi_\nu(f(T)) = \Phi_{\nu 1}(f_1(T)) + \ldots + \Phi_{\nu n}(f_n(T)) = 0 \qquad (1)$$

for all $\nu \in N$. Assume, as we may, that $f_1(T)$ is nonconstant and has a term $bT^s$ with $b \neq 0$. Write $s = up^m$, $(u, p) = 1$, and for each $1 \leq i \leq n$ define

$g_i(T) :=$ the sum of all terms of $f_i(T)$ of degree $up^h$ for any $h \geq 0$

and set $g_i(T) := 0$ if no terms of the form $cT^{up^h}$ ($c \in k$) is present in $f_i(T)$. It follows from (1) above, then, that $\Phi_{\nu 1}(g_1(T)) + \ldots + \Phi_{\nu n}(g_n(T)) = 0$ for all $\nu \in N$ by the manner the $g_i(T)$ are built. One can now construct

$\phi: G_a \to G$ by defining $\phi(R)(y) = (g_1(y^{1/u}), \ldots, g_n(y^{1/u})) \in G(R)$ for all $y \in G_a(R)$.

Proof of 4.3.1. We now prove (i) $\Rightarrow$ (iii) of the theorem. Suppose given a nonconstant k-morhpism $\psi : \mathbb{A}^1 \to \overline{G}$ and extend the ground field to the separable closure $K = k_s$ to obtain $\psi_s = \psi \otimes K : \mathbb{A}^1_K \to \overline{G}_K$. If the image of $\psi_s$ is not contained in the center of $G_K$, take a point $g \in G_K(K)$ not centralizing the image and replace $\psi_s$ by the morphism $x \mapsto \psi_s(x) g\, \psi_s(x)^{-1} g^{-1}$. The image of the new nonconstant $k_s$-morphism is contained in $[G_K , G_K] = [G , G] \otimes K$. By repeating this process as long as necessary, one has at hand a nonconstant $k_s$-morphism $\phi_s : \mathbb{A}^1_K \to \overline{G}_K$ whose image is contained in the center of $G_K$. Next, if the image of $\phi_s$ is not contained in the kernel of the p-th power endomorphism $p \cdot id : x \mapsto x \,.....\, x$ (p factors) on $G$, compose $\phi_s$ with that endomorphism and repeat the process until the image of $\mathbb{A}^1_K$ is killed by the p-th power endomorphism. In all, we have a nonconstant $k_s$-morphism $\eta_s : \mathbb{A}^1_K \to Z_K \subseteq G_K$ with a suitable central k-closed subgroup $Z \subseteq G$ of exponent p. By 4.5, we may assume $\eta_s \in \mathrm{Hom}_{K\text{-}gr}((G_a)_K, Z_K)$ and may apply 4.4 : Since the $k_s$-module $\mathrm{Hom}_{K\text{-}gr}((G_a)_K, Z_K) \neq \{0\}$, its k-form $\mathrm{Hom}_{k\text{-}gr}(G_a , Z)$ cannot be reduced to $\{0\}$ and there is hence a nonconstant k-homomorphism $\eta : G_a \to Z \subseteq G$.

4.6. DEFINITION. A unipotent algebraic k-group is said to be k-wound if any one of the equivalent conditions of 4.3.1. holds for it.

4.7. PROPOSITION. Let $G$ be a connected one-dimensional unipotent k-group. Then, the following are equivalent :

(i) $G$ is k-wound ;

(ii) $G$ is a nontrivial k-form of $G_a$ .

(iii) $\overline{G}$ is a nontrivial k-form of $\mathbb{A}^1$ .

This is an easy corollary to 4.2, 4.3.1, and to the fact that $\mathbb{A}^1$ admits a unique k-group structure up to the choice of the origin [2; §3].

## 5. The question of commutativity for two-dimensional unipotent groups

*In this section (§5), the ground field k is assumed to be imperfect of characteristic* p.

5.0. In spite of what we saw in §3 (see, esp., 3.7), the state of our knowledge on two-dimensional unipotent k-groups leaves much to be desired. In particular, the study of those which are k-wound presents considerable difficulty for obvious reasons. Among other things, there has been a conjecture in some quarters to the effect that every k-wound connected two-dimensional unipotent k-group is commutative. In the present section, we show that the conjecture itself is false but under mild, additional hypotheses all such groups turn out to be commutative (see 5.8 to 5.8.4). We start out with some explicit calculations of homomorphisms between k-groups of Russell type. The calculations pave the way for later results in §5, but they are also of some independent interest.

5.1. Let G be a k-form of a vector group discussed in §2. By *the height of* G, denoted ht(G), we shall mean the least $n \geqq 0$ such that $G^{(p^n)}$ is k-isomorphic to the vector group. By 2.5, this is equivalent to saying that ht(G) is the least $n \geqq 0$ such that $G \otimes k^{p^{-n}}$ is $k^{p^{-n}}$-isomorphic to a vector group over $k^{p^{-n}}$. In the present paper, we shall actually use this concept only when $G$ is a k-form of $G_a$, i.e., when $G$ is k-isomorphic to either $G_a$ or a k-group of Russell type.

5.2. Let $G=\underline{D}(M(m,\alpha))$ be a k-goup of Russell type, where $\alpha \in k[F]$, $m > 0$ and $(m,\alpha)$ is admissible in the sense of 2.7. Thus, $ht(G)=m$ by 2.5. Write $\mathcal{O}(G)=k[x,y]$, $Ⓔ=k[F]$ and (by a slight abuse of notation) $\underline{M}(G)=Hom_{k\text{-}gr}(G, G_a)=Ⓔx+Ⓔy$, $\alpha x=F^m y$. Write $\alpha=a_0+a_1F+....+a_rF^r$, and let $k'$ be a field containing $k(a_1^{p^{-m}},\dots, a_r^{p^{-m}})$. Let as above $Ⓔ=k[F]$ and let $Ⓔ'=k'[F]$. Consider the $Ⓔ'$-module $\underline{M}'(G):=Hom_{k'\text{-}gr}(G_{k'},(G_a)_{k'})$. It is easy to verify that $\underline{M}'(G)$ is $Ⓔ'$-isomorphic to $k'\otimes \underline{M}(G)$ made into $Ⓔ'$-module *via* $F(\lambda\otimes m):=\lambda^p\otimes Fm$ for $\lambda \in k'$, $m\in \underline{M}(G)$.

5.2.1. PROPOSITION. *Through the mapping* $m \mapsto 1\otimes m$ *one can identify* $\underline{M}(G)$ *as an E-submodule of the* $Ⓔ'$*-module* $\underline{M}'(G)=k'\otimes \underline{M}(G)$ *which is* $Ⓔ'$*-free of rank one. A free base* $\{z\}$ *of* $\underline{M}'(G)$ *may be chosen so that* $\underline{M}'(G)=Ⓔ'z$, $x=F^m z$ *and* $y=\alpha^{(-m)}z$, *where* $\underline{M}(G)=Ⓔx+Ⓔy$, $\alpha x=F^m y$ *as above*. (see 2.4 for the notation $\alpha^{(-m)}$.)

*Proof* (Sketch). Let $\widetilde{Ⓔ}'$ be the left total quotient ring of $Ⓔ'$ and imbed the torsion-free left $Ⓔ'$-module $\underline{M}'(G)$ into $\widetilde{Ⓔ}'\otimes_{Ⓔ'}\underline{M}'(G)$ in the standard fashion. Having already imbedded $\underline{M}(G)$ inside $\underline{M}'(G)$ as indicated above, we perform calculations in $\widetilde{Ⓔ}'\otimes_{Ⓔ'}\underline{M}'(G)$ to find

$$\begin{cases} 1 \otimes x = F^{m}(F^{-m} \otimes x) \\ 1 \otimes y = F^{-m}\alpha \otimes x = \alpha^{(-m)}(F^{-m} \otimes x) \end{cases} \tag{1}$$

whence $\underline{M}'(G) \subseteq \textcircled{E}'(F^{-m} \otimes x) \subseteq \textcircled{E}' \otimes \underline{M}'(G)$. But, by virtue of the right divison algorithm in $\textcircled{E}'$, the left ideal $J' = \{\phi' \in \textcircled{E}' : \phi'(F^{-m} \otimes x) \in \underline{M}'(G)\} \subseteq \textcircled{E}'$ is principal, so that $J' = \textcircled{E}'\gamma'$ with $\gamma' \in \textcircled{E}'$. By (1) we have $F^{m} \in J'$ and $\alpha^{(-m)} \in J'$ ; since $F^{m}$ is purely inseparable and $\alpha^{(-m)}$ is separable as $k'$-morphisms, $\gamma'$ must be a nonzero constant $\in k'$ and hence $J' = \textcircled{E}'$. Therefore, $\underline{M}'(G) = \textcircled{E}'(F^{-m} \otimes x) = \textcircled{E}'$-free of rank one, and $F^{-m} \otimes x$ serves as our $z$. For more details, see DG-IV, §3, n°6.

5.3. Let $G_1 = \underline{D}(M(m,\alpha))$, $G_2 = \underline{D}(M(n,\beta))$ be either $G_a$ or $k$-groups of Russell type, where $(m,\alpha)$, $(n,\beta)$ are admissible pairs and $\alpha = a_0 + a_1F + \ldots + a_rF^{r}$, $\beta = b_0 + b_1F + \ldots + b_sF^{s}$. Thus, $m = ht(G_1)$ and $n = ht(G_2)$.

5.3.1. THEOREM. *With these notations, we have:*

(i) *If* $m > n$ *then* $Hom_{k\text{-}gr}(G_2, G_1) = \{0\}$ ; *and*

(ii) *If* $1 \leq m \leq n$, *then* $Hom_{k\text{-}gr}(G_2, G_1)$ *is isomorphic as additive group to the group of all elements* $\phi \in \textcircled{E}' = k'[F]$ *with* $k' = k^{p^{-n}}$ *such that*

$$F^{m}\phi = \phi^{(m)}F^{m} \in \textcircled{E}F^{n} + \textcircled{E}\beta^{(-n)} \quad \text{and} \quad \alpha^{(-m)}\phi \in \textcircled{E}F^{n} + \textcircled{E}\beta^{(-n)} \tag{2}$$

<u>where</u> Ⓔ = k[F]. (See 2.4 for the notations $\phi^{(m)}$, $\alpha^{(-m)}$.)

<u>Proof</u>. (i) This is clear since $\mathrm{Hom}_{k\text{-gr}}(G_2, G_1) = \mathrm{Hom}_{Ⓔ}(\underline{M}(G_1), \underline{M}(G_2))$ by 1.5.1 and this last is reduced to {0} by 2.8.1. (This can also be proven by applying the Frobenius functor n times to the situation.)

(ii) By means of 5.2.1, one way write: $\underline{M}(G_1) = Ⓔx+Ⓔy$, $\alpha x = F^m y$, $\underline{M}'(G_1) = Ⓔ'z$ with $x = F^m z$, $y = \alpha^{(-m)}z$; and $\underline{M}(G_2) = Ⓔu+Ⓔt$, $\beta u = F^n t$, $\underline{M}'(G_2) = Ⓔ'w$ with $u = F^n w$, $t = \beta^{(-n)}w$. Given an Ⓔ-module homomorphism $\underline{M}(G_1) \to \underline{M}(G_2)$, it corresponds uniquely to an Ⓔ'-module homomorphism $\underline{M}'(G_1) \to \underline{M}'(G_2)$ mapping $\underline{M}(G_1)$ into $\underline{M}(G_2)$, and <u>vice versa</u> by the restriction of homomorphism. Since an Ⓔ'-module homomorphism $\underline{M}'(G_1) \to \underline{M}'(G_2)$ is completely determined by $\phi \in Ⓔ'$ such that $z \longmapsto \phi w$, the set of all Ⓔ-module homomorphisms $\underline{M}(G_1) \to \underline{M}(G_2)$ is in one-to-one correspondence with the set of $\phi \in Ⓔ'$ such that $F^m\phi w \in \underline{M}(G_2) = (ⒺF^n + Ⓔ\beta^{(-n)})w$ and $\alpha^{(-m)}\phi w \in \underline{M}(G_2)$.

5.4. Let $G_1 = \underline{D}(M(n,\alpha))$, $G_2 = \underline{D}(M(n,\beta))$ be k-groups of the same height n. As before, let us write $\underline{M}(G_1) = Ⓔx+Ⓔy$, $x = F^n z$, $y = \alpha^{(-n)}z$ and $\underline{M}(G_2) = Ⓔu+Ⓔt$, $u = F^n w$, $t = \beta^{(-n)}w$. The following proposition comes handy in pair with 5.3. :

5.4.1. PROPOSITION. <u>With the notations as above, we have</u>:

(i) <u>If</u> $n \geq 1$, <u>then</u> $\mathrm{Hom}_{k\text{-gr}}(G_2, G_1)$ <u>is isomorphic</u>

as additive group to the group of all elements $\phi \in \textcircled{E} = k[F]$ for which there exist $\xi \in \textcircled{E}$ and $\eta \in \textcircled{E}$ such that

$$\alpha\phi = \xi^{(n)}F^n + \eta^{(n)}\beta \ . \tag{3}$$

Such a triple $(\phi, \xi, \eta)$ gives the homomorphism $M(G_1) \to \underline{M}(G_2)$ defined by $x \longmapsto \phi u$, $y \longmapsto \xi u + \eta t$.

(ii) If $n > 1$, then the mapping which sends the triple $(\phi, \xi, \eta)$ subject to (3) to the triple $(\phi, \xi^{(1)}F, \eta^{(1)})$ realizes the Frobenius homomorphism $f \in \mathrm{Hom}_{k\text{-}gr}(G_2, G_1) \longmapsto f^{(p)} \in \mathrm{Hom}_{k\text{-}gr}(G_2^{(p)}, G_1^{(p)})$.

These are immediate consequences of 5.3.1-(ii) and elementary calculations. We omit the details. Note only that the homomorphism in (ii) above is injective and $f$ is a monomorphism if and only if $f^{(p)}$ is such. Note also that $f \neq 0$ is always an epimorphism.

We list a simple fact (already used in 2.8.1) as a lemma here for the sake of easy reference:

5.5. LEMMA. Let $\alpha = \sum_0^n a_i F^i \in \textcircled{E} = k[F]$ be such that at least one of $a_1, a_2, \ldots$ be a non-p-th power in $k$. Then, $\alpha\phi = F\psi$ for $\phi \in \textcircled{E}$, $\psi \in \textcircled{E}$ implies $\phi = \psi = 0$. (Proof is omitted.)

Note that this lemma shows among other things that the $\eta$ appearing in (3) of 5.4.1 is never zero for any nonzero $k$-homomorphism $G_2 \to G_1$.

5.6. If $G = \underline{\underline{D}}(M(n,\alpha))$ is a k-group of Russell type, an overfield $k' \supseteq k$ satisfies $G \otimes k' \simeq (G_a)_{k'}$ if and only if $k' \supseteq k(a_1^{p^{-n}}, \ldots, a_r^{p^{-n}})$ — see [11; Lemma 1.3, p.529]. The field $k(a_1^{p^{-n}}, \ldots, a_r^{p^{-n}})$ is called <u>the minimum splitting field of</u> G.

5.6.1. PROPOSITION. <u>Let G be a k-wound k-group of Russell type, and let K be a field containing k but not containing the minimum splitting field of G. Then, there exists a power of p, $d = d(K) = p^{\nu}$ with $\nu \geq 1$, such that</u>

$$\mathrm{Aut}_{K\text{-gr}}(G_K) = \{x \in K : x^{d-1} = 1\}, \quad \underline{\text{and}}$$

$$\mathrm{End}_{K\text{-gr}}(G_K) = \{y \in K : y^d = y\}.$$

<u>Moreover, for any fields $k \subseteq K_1 \subseteq K_2$ none of which containing the minimum splitting field of G, $d(K_2)$ is a positve integral power of $d(K_1)$ and a fortiori $d(K_1) \leq d(K_2)$; the last becomes an equality whenever $K_2$ is separable over $K_1$.</u>

The proposition duplicates Russell's result [11; Th. 3.1, p.536]. Therefore, we shall merely sketch its proof, though our method is somewhat different from Russell's.

<u>Proof</u> (Sketch). By 5.4.1-(ii) and remarks following it, it is enough to consider the case ht(G)=1. Let, therefore, $G = \underline{\underline{D}}(M(1,\alpha))$ and look at $G_K$. The defining equation for

$G_K$ is $y^p = a_0x+a_1x^p+\dots+a_rx^{p^r}$, but one can quickly ascertain that one may assume with no loss of generality that $a_0=1$, $a_r \notin K^p$. That being assumed, let us look for $\phi \in K[F]$ satisfying

$$\alpha\phi = \xi^{(1)}F + \eta^{(1)}\alpha \tag{4}$$

for some $\xi \in K[F]$, $\eta \in K[F]$. We may assume $\eta = y \in K$ by 5.4.1-(ii). Then $\phi$, too, must be an element of $K$ because $a_r \notin K^p$, so write $\phi=c$. Then, $\alpha\phi - \eta^{(1)}\alpha = \alpha c - y^p\alpha = \xi^{(1)}F$ and $a_0=1$, whence $c=y^p$. Now we have $\alpha y^p - y^p\alpha = x_1^pF+\dots x_r^pF^r$ $=\xi^{(1)}F$ for some $x_1,\dots,x_r \in K$ and $y \in K$. Thus, our $y$ must satisfy the condition

$$a_i(y^p)^{p^i} - a_iy^p = a_i(y^{p^i} - y)^p \in K^p \tag{5}$$

for all $1 \leq i \leq r$, or equivalently $y^{p^i}-y=0$ for every $i$ for which $a_i \notin K^p$. Conversely, each $y \in K$ satisfying (5) gives a solution of (4) if one sets $\phi=y^p$. This shows what we asserted above.

5.6.2. REMARK. Consider the automorphism functor $\underline{\mathrm{Aut}}_G : \boxed{\mathrm{Alg}}_k \longrightarrow \boxed{\mathrm{Gr}}$ given by $\underline{\mathrm{Aut}}_G(R) := \underline{\mathrm{Aut}}_{R\text{-}gr}(G_R)$, where as before $G$ is a k-group of Russell type. If $k'$ is a splitting field of $G$, Then $\underline{\mathrm{Aut}}_G \otimes k' = \underline{\mathrm{Aut}}_{(G_a)_{k'}} =$ $\underline{\mathrm{Aut}}_{G_a} \otimes k'$, whence $\underline{\mathrm{Aut}}_G$ is a k-form of $\underline{\mathrm{Aut}}_{G_a}$. It is easy to show that $\underline{\mathrm{Aut}}_{G_a}$ is non-representable, and therefore $\underline{\mathrm{Aut}}_G$ is not, either. Notice, however, that $\underline{\mathrm{Aut}}_{G_a}(K) = K^{\times} = G_m(K)$ for any field $K \supseteq k$.

5.6.3. EXAMPLE. Let $a \in k$, $a \notin k^p$, and let $G_1 = \underline{\underline{D}}(M(1, 1-aF))$, $G_2 = \underline{\underline{D}}(M(1, 1+aF^2))$, both of height 1. We compute $\mathrm{Hom}_{k\text{-}gr}(G_2, G_1)$ by means of 5.4.1-(i). As in the proof of 5.6.1, we look for all $\phi \in k[F]$ for which there are $\xi \in k[F]$ and $c \in k$ such that $\alpha\phi - c^p\beta = \xi^{(1)}F$, where $\alpha = 1-aF$, $\beta = 1+aF^2$. Immediately we learn that $\phi$ is of the form $\phi = c^p + xF$ with $x \in k$. Further calculations show that $x = -c$ and $c$ is subject to the condition that $c + ac^{p^2} \in k^p$. One then concludes that $\mathrm{Hom}_{k\text{-}gr}(G_2, G_1)$ is isomorphic as additive group to the k-rational points $G_2(k)$ of $G_2$. The fact stands valid if $k$ is replaced by any other field not containing $a^{p^{-1}}$, in particular by $k_s$. Thus, $\mathrm{card}\,[\mathrm{Hom}_{k_s\text{-}gr}((G_2)_{k_s}, (G_1)_{k_s})] = \infty$ in this example. (Cf. 5.10 below.)

5.7. We now turn to the main question of this section, i.e., the question of commutativity for two-dimensional unipotent groups. Let $G$ be a k-wound connected unipotent algebraic k-group of dimension 2. If $G$ is non-commutative, there arises a central exact sequence $0 \to G_1 \to G \to G_2 \to 0$ where $G_1 = [G, G]$ and $G_2 = G/G_1$. Clearly, both $G_1$ and $G_2$ are k-forms of $G_a$ and $G_1$ is k-wound; hence so is

$G_2$ by virtue of 3.7. Thus, to study the groups like $G$ above, one must look at central extensions of a k-group of Russell type by another such group. By doing so, we shall prove a number of sufficient conditions for the absence of noncommutative central extensions of the envisaged type.

5.8. THEOREM. _Let_ $G_1$ _be a commutative_ k-_group scheme and let_ $G_2$ _be a_ k-_group of dimension one (necessarily commutative). Suppose that_

$$\mathrm{card}(\mathrm{Hom}_{k_s\text{-gr}}((G_2)_{k_s}, (G_1)_{k_s})) < \infty. \tag{6}$$

_Then_, $\mathrm{Ext}_{\mathrm{cent}}(G_2, G_1) = \mathrm{Ext}_{\mathrm{com}}(G_2, G_1)$, i.e., _all central extensions of_ $G_2$ _by_ $G_1$ _are commutative_.

_Proof_. By 3.2-(i), it suffices to show that every antisymmetric biadditive function $G_2 \times G_2 \to G_1$ is zero. So, suppose given a nonzero biadditive function $B : G_2 \times G_2 \to G_1$; by extending the base field to $k_s$ we still have a nonzero biadditive function on $(G_2)_{k_s}$, and therefore we shall assume without loss that $k=k_s$ already. Since $B$ is nonzero, there exists a point $(a,b) \in (G_2 \times G_2)(k)$ such that $B(a, b) \neq 0$. Consequently, the function $B(a, -)$ is a nonzero k-homomorphism $G_2 \to G_1$, hence $B(a, x) = 0$ for only finitely many $x \in G_2(k)$. On the other hand, map each $x \in G_2(k)$ onto $B(-, x) \in \mathrm{Hom}_{k\text{-gr}}(G_2, G_1)$ to obtain a homomorphism $G_2(k)$ (infinite) $\longrightarrow$ $\mathrm{Hom}_{k\text{-gr}}(G_2, G_1)$ (finite); its kernel is infinite; therefore, for infinitely many $x \in G_2(k)$, we must have $B(-, x) = 0$ as function and hence a fortiori $B(a, x) = 0$ for infinitely many $x \in G_2(k)$. This is a contradiction. Q.E.D.

5.8.1. COROLLARY. Let $G_1$ and $G_2$ be one-dimensional unipotent k-groups, and suppose that either (i) $\mathrm{ht}(G_1) > \mathrm{ht}(G_2)$ or (ii) $\mathrm{card}(\mathrm{Hom}_{k_s\text{-gr}}((G_1)_{k_s}, (G_2)_{k_s})) = \infty$ and $G_2$ is k-wound. Then, $\mathrm{Ext}_{cent}(G_2, G_1) = \mathrm{Ext}_{com}(G_2, G_1)$ holds.

Proof. Under the condition (i), the assertion follows obviously from 5.3.1-(i). Now assume (ii). The conclusion comes then from 5.8 and from the next lemma.

5.8.2. LEMMA. Let $G_1$ and $G_2$ be one-dimensional unipotent k-groups, and suppose that $\mathrm{card}(\mathrm{Hom}_{k\text{-gr}}(G_1, G_2)) = \infty$ and $G_2$ is k-wound. Then, $\mathrm{Hom}_{k\text{-gr}}(G_2, G_1) = \{0\}$.

Proof. Assume that there were a nonzero k-homomorphism $\phi : G_2 \to G_1$, which would then be an epimorphism clearly. Then, for all $\xi \in \mathrm{Hom}_{k\text{-gr}}(G_1, G_2)$ the k-homomorphisms $\xi\phi \in \mathrm{End}_{k\text{-gr}}(G_2)$ would have to be mutually distinct. Since the first Hom set is infinite whereas the second is finite by 5.6.1, we have gotten a contradiction.

5.8.3. COROLLARY. Let G be a k-wound one-dimensional unipotent k-group. Then, $\mathrm{Ext}_{cent}(G, G)=\mathrm{Ext}_{com}(G, G)$.

This is obvious in view of 5.6.1 and 5.8.

5.8.4. COROLLARY. Let E be a k-wound connected two-dimensional noncommuative unipotent k-group. Then, $\mathrm{card}(\mathrm{End}_{k_s\text{-}gr}(E_{k_s})) = \infty$.

Proof. There is no loss in making the working hypothesis that $k=k_s$. By the noncommutativity, we can construct a central exact sequence $1 \to G_1 \to E \to G_2 \to 1$, where $G_1=[E, E]$, $G_2$ are one-dimensional k-wound k-groups and $ht(G_1) \leqq ht(G_2)$ by virture of 5.8.1-(i). Moreover, by 5.8, $\mathrm{Hom}_{k\text{-}gr}(G_2, G_1)$ is infinite. Then, as $E \to G_2$ is an epimorphism there arises an inclusion map $\mathrm{Hom}_{k\text{-}gr}(G_2, G_1) \longrightarrow \mathrm{Hom}_{k\text{-}gr}(E, G_1)$ and consequently $\mathrm{Hom}_{k\text{-}gr}(E, G_1)$ is infinite. But $G_1 \to E$ is a monomorphism, so that another inclusion $\mathrm{Hom}_{k\text{-}gr}(E, G_1) \rightarrow \mathrm{Hom}_{k\text{-}gr}(E, E)$ obtains, which shows that the last Hom set is infinite.

5.9. Let E be a k-wound connected two-dimensional unipotent noncommutative k-group. We have seen that E may be written in the form of a central extension of k-forms of $G_a$, $1 \to G_1 \to E \to G_2 \to 1$, with $1 \leqq ht(G_1) \leqq ht(G_2)$. By 5.8, we know that $\mathrm{Hom}_{k_s\text{-}gr}((G_2)_{k_s}, (G_1)_{k_s})$ is infinite. We conjecture that the converse is valid. To wit: If

$\mathrm{Ext}_{\mathrm{cent}}(G_2, G_1) = \mathrm{Ext}_{\mathrm{com}}(G_2, G_1)$, then $\mathrm{card}[\mathrm{Hom}_{k_s\text{-gr}}((G_2)_{k_s}, (G_1)_{k_s})] < \infty$.(*)

5.10. Example. We present what is probably the first example of a noncommutative k-wound connected unipotent k-group of dimension 2. It was first discovered for $m=n=1$ (see below) by Shizuo Endo.

Let $a \in k$, $a \notin k^p$, and let $m$, $n$ be integers subject to $0 \leq m \leq n$. Let $G_1 = \underline{\underline{D}}(M(m, 1-aF^m))$, $G_2 = \underline{\underline{D}}(M(n, 1+aF^{2m}))$. We produce a nonzero antisymmetric biadditive function $G_2 \times G_2 \to G_1$ as follows: Writing $\overline{G_1} = \mathrm{Spec}\ k[x,y]$, $y^{p^m} = x - ax^{p^m}$ and $\overline{G_2} = \mathrm{Spec}\ k[u, t]$, $t^{p^n} = u + au^{p^{2m}}$, we define a function $k[x, y] \longrightarrow k[u, t] \otimes k[u, t]$ through the assignments $x \longmapsto u^{p^m} \otimes u - u \otimes u^{p^m}$ and $y \longmapsto u \otimes t^{p^{n-m}} - t^{p^{n-m}} \otimes u$. It is mechanical to verify that this gives an antisymmetric biadditive function $f : G_2 \times G_2 \to G_1$. Then, using $f$ as a 2-cocycle, construct a central extension $1 \to G_1 \to G_2 \times_f G_1 \to G_2 \to 1$. The middle term is the desired noncommutative k-wound unipotent k-group. Recall that we calculated $\mathrm{Hom}_{k\text{-gr}}(G_2, G_1)$ in case $m = n = 1$ already — cf. 5.6.3 and 5.9 above.

---

(*) In early 1974 a counter-example to this conjecture was constructed by Mr. Tsutomu Oshima, a graduate student at Tokyo Metropolitan University.

# Part II

## 6. Forms of the affine line and geometry of the groups of Russell type

In this section (§6), k is an imperfect field of characteristic p.

6.0. We investigate in the present section the k-forms of $\mathbb{A}^1$. After some preparations on purely inseparable descent and derivations, we firstly present in 6.5 a necessary and sufficient condition in order for a k-form of $\mathbb{A}^1$ to be trivial. Next, in 6.7, we consider the function field of such a form and give a characterization 6.7.9 of all k-rational k-forms of $\mathbb{A}^1$. In particular, we determine all k-rational k-groups of Russell type in 6.9.2. Thirdly, we classify completely all k-forms of $\mathbb{A}^1$ whose arithmetic genera are either 0 or 1 and which carry k-rational points (see 6.8.1 and 6.8.3). Lastly, we make out some explicit calculations and determine the Picard groups of the underlying k-schemes of certain Russell type k-groups (see 6.10 and the paragraphs that follow it — especially 6.10.1, 6.11.1 and 6.12.2). We have also appended a remark in 6.13 concerning $\mathrm{Ext}_{\mathrm{cent}}(G,G_m)$ for k-groups G of Russell type.

Notations used only in this section (§6) include the following: For a k-scheme X, Pic(X): = the group of isomorphism classes of invertible sheaves on X; for a k-algebra A,

$C(A)$: = the divisor class group of $A$ as in [12]; and $\$^1$: = Spec $k[T,T^{-1}] = \bar{G}_m = (\mathbb{A}^1$ - {k-rational point}). Also, Pic and Aut are used, for which see 6.7 and 6.9, respectively.

Our thanks are due to Mike Artin for pointing out some useful facts relevant to this section. In particular, we owe 6.7.3. and a proof of 6.7.7 to him.

We begin by some remarks and examples of purely inseparable forms of $\mathbb{A}^1$ and $\mathbb{A}^2$.

6.1. LEMMA.* *Let* $k' = k(\lambda)$ *be a purely inseparable simple extension of* $k$ *with* $\lambda^p \in k$, $\lambda \notin k$. *Let* $A'$ *be a* $k'$*-algebra. Then: Given a* $k$*-subalgebra* $A$ *such that* $k' \otimes A \simeq A'$, *there is a unique* $A$*-derivation* $D$ *on* $A'$ *such that* $D^p = 0$ *and* $D(\lambda) = 1$. *Conversely, for each* $k$*-derivation* $D$ *on* $A'$ *with* $D^p = 0$ *and* $D(\lambda) = 1$, *the* $k$*-subalgebra* $A = A'^D$ *(= the ring of* $D$*-constants) satisfies* $A' \simeq k' \otimes A$.

*Proof*. Let $A$ be given as above. Let $d$ be the unique $k$-derivation on $k'$ such that $d(\lambda) = 1$. Let $D = d \otimes 1$, so that $D(\sum_i \lambda_i \otimes a_i) = \sum_i d(\lambda_i) \otimes a_i$. It is obvious that $D^p = 0$, $D(\lambda) = 1$ and $A \subseteq A'^D$. To prove $A'^D \subseteq A$, write any element $a'$ of $A'$ in the form $a' = \sum_{i=0}^{p-1} a_i\lambda^i$ with $a_i \in A$. Then $D(\sum_{i=0}^{p-1} a_i\lambda^i) = \sum_{i=0}^{p-1} ia_i\lambda^{i-1}$. Thus $a_i = 0$ for $i=1,\cdots,$

---

* This lemma also follows easily from Cor.7.10 in the next section.

$p - 1$. Hence $a' = a_0 \in A$.

Conversely let $D$ be given as before. It is easy to see that $A = A'^D$ is a k-subalgebra of $A'$. Let $a'$ be any element of $A'$. Since $D^p = 0$, there is an integer $n$ $(0 \leq n < p)$ such that $D^n(a') \neq 0$ and $D^i(a') = 0$ for all $i > n$. Then $D^n(a') \in A$, and $D^n(a' - \frac{1}{n!} D^n(a')\lambda^n) = 0$. By induction on $n$, we know that any element $a'$ of $A'$ can be written in the form $\sum_{i=0}^{p-1} a_i \lambda^i$. The expression is unique. In fact, let $\sum_{i=0}^{p-1} a_i \lambda^i = 0$. Let $n$ be the integer such that $a_n \neq 0$ and $a_i = 0$ for all $i > n$. Then $D^n(\sum_{i=0}^{p-1} a_i \lambda^i) = n! a_n = 0$. Thus $a_n = 0$. This is a contradiction. Hence $a_0 = \cdots = a_{p-1} = 0$.

6.2. This lemma is useful in constructing purely inseparable forms of $\mathbb{A}^1$ or $\mathbb{A}^2$. We shall give two examples, the first one of which is a reconstruction of a one-dimensional unipotent group of Russell type and the other is a non-trivial purely inseparable form of $\mathbb{A}^2$.

6.2.1. Example. Let $k' = k(\lambda)$ with $\lambda^p = \alpha \in k$, $\lambda \notin k$. Let $A' = k'[t]$ = the polynomial ring over $k'$ in one variable $t$. Let $D$ be a k-derivation on $A'$ defined by $D(\lambda) = 1$, $D^p = 0$ and $D(t) = t^p$. Then $A = A'^D = k[t^p, t+\lambda t^p]$. Write $y = t + \lambda t^p$ and $x = t^p$. Then $A = k[x,y]/(y^p = x+\alpha x^p)$.

6.2.2. <u>Example</u>. Let $p = 2$ and let $k' = k(\lambda)$ with $\lambda^2 = \alpha \in k$, $\lambda \notin k$. Let $A' = k'[x,y]$ be the polynomial ring over $k'$ in two variables $x$ and $y$. Define a $k$-derivation $D$ on $A'$ by $D(\lambda)=1$, $D(x)=xy$ and $D(y)=y^2$. It is easy to see that $D^2 = 0$ on $A'$. The ring of $D$-constants $A = A'^D$ is $k[x^2,y^2,xy,x+\lambda xy,y+\lambda y^2]$. We shall show that $A$ is not $k$-isomorphic to a polynomial ring over $k$ of two indeterminates. For this it suffices to show that $A$ is not a unique factorization domain. Let $t=y^2$, $s=xy$, $u=x+\lambda xy$ and $v=y+\lambda y^2$. When their decomposition in $k'[x,y]$ is considered, it is clear that $t$, $s$, $u$, $v$ are all irreducible elements in $A$. However $tu = sv$. Hence $A$ is not a unique factorization domain.

6.3. We shall next look for a necessary and sufficient condition for a purely inseparable form of $\mathbb{A}^1$ or $\mathbb{S}^1$ to be trivial. Our main tool is the following Samuel's Theorem, which we shall recall below for the reader's convenience and of which we shall make a slight generalization.

Let $A'$ be a Krull domain of characteristic $p \neq 0$ and let $K'$ be the quotient field of $A'$. Assume that there is given a derivation $D \neq 0$ on $K'$ such that $D(A') \subseteq A'$. Let $K = K'^D$ and let $A = A' \cap K$. $A$ is then a Krull domain. Denote by $C(A')$ and $C(A)$ the divisor class groups of $A'$ and $A$, respectively. An element of $C(A)$ is written in the form $\sum n_P P$, where $P$ runs over all prime ideals of height

1 of A and $n_P$ is an integer such that $n_P = 0$ for almost all P. Since $A'^p \subseteq A$, above each P there is one and only one prime ideal $P' \subseteq A'$ of height 1, whose ramification index $e_{P'}$ is either 1 or p. Assigning $\sum n_P e_{P'} P'$ to $\sum n_P P$, we have the canonical map $j: C(A) \to C(A')$.

To describe Ker(j), we shall introduce the abelian group L consisting of logarithmic derivatives $D(z)/z$ with $z \in K'^{\times}$ and $D(z)/z \in A'$. Let $L_0$ be the subgroup of L consisting of $D(u')/u'$ with $u' \in A'^{\times}$. With these notations, Samuel's Theorem states:

6.3.1. LEMMA. (P. Samuel [12; Theorem 3.2, p.62]). *With the notations and assumptions as above, assume moreover that* D(A') *is not contained in any prime ideal of height* 1 *in* A'. *Then we have an exact sequence.*

$$0 \to L/L_0 \xrightarrow{i} C(A) \xrightarrow{j} C(A').$$

*The map* i *sends* $D(z)/z$ (modulo $L_0$) *to* $\sum (v_{P'}(z)/e_{P'}) \cdot (P' \cap A)$ *where* $v_{P'}$ *is the discrete valuation on* K' *corresponding to* P'. *The map* j *is surjective if and only if the ramification index* $e_{P'}$ *is* 1 *for all prime ideals* P' *of height* 1 *of* A'.

6.4. To generalize Samuel's Theorem, consider the following situation:

Let k be a field of characteristic $p > 0$ and G a <u>finite</u> connected k-group scheme. By G finite we mean that the algebra $O(G)$ is finite dimensional. Put $rk(G) = [O(G):k]$. Let N be the <u>height of</u> G, defined as the smallest integer N such that the N-times iterated Frobenius map $F^N: G \longrightarrow G^{(p^N)}$ is trivial. Let A' be a Krull domain k-algebra and let K' be the quotient field of A'. Assume that the k-group scheme G acts <u>freely</u> on the affine k-scheme $X' = \mathrm{Spec}(A')$ on the right via

$$u: X' \times G \longrightarrow X'.$$

Let $\rho: A' \longrightarrow A' \otimes O(G)$ be the associated coaction. (The reader can consult §7 for the concepts of free action and coaction.) If we put $A = A'^G$, then it follows from DG-III, §2, n°6 that A' is a finite projective A-module of rank $rk(G)$. Since $f^{p^N} \in k$ for any $f \in O(G)$, it follows that $a'^{p^N} \in A$ for any $a' \in A'$. In particular every element of K' is of the form b/a with $a \in A - (0)$ and $b \in A'$. Therefore the coaction $\rho$ can be uniquely extended to a coaction

$$\rho: K' \longrightarrow K' \otimes O(G)$$

which is free too, clearly. Let K be the quotient field of A. Since $K = K'^G$, we have $K \cap A' = A$. Hence A is a Krull domain and the prime ideals of height 1 of A correspond bijectively with those of A' through $P = P' \cap A \longleftrightarrow P'$. The ramification index $e_{P'}$ of $P'|P$ divides $p^N$ clearly. The map $j: C(A) \longrightarrow C(A')$ is defined in the same fashion as above. Put

$$\chi: K'^{\times} \longrightarrow (K' \otimes O(G))^{\times},\ z \longmapsto \chi(z) = (z \otimes 1)^{-1}\rho(z).$$

Then $\chi$ is a homomorphism of abelian groups and $\mathrm{Ker}(\chi) = K^{\times}$ clearly. Hence $\chi(K'^{\times}) \simeq K'^{\times}/K^{\times}$ is an abelian group of exponent $p^N$. Therefore

$$L := \chi(K'^{\times}) \cap (A' \otimes O(G))$$

is a subgroup of $\chi(K'^{\times})$ and contains the subgroup

$$L_0 := \chi(A'^{\times}).$$

6.4.1. The multiplicative groups $L$, $L_0$ we defined just now comprise as special cases Samuel's groups $L$, $L_0$ of logarithmic derivatives introduced in 6.3 above. To see that fact, consider the case when $G = \alpha_p \times \ldots \times \alpha_p$ (r factors). Write $O(G) = k[T_1,\ldots, T_r]/(T_1^p,\ldots, T_r^p) = k[t_1,\ldots, t_r]$. Let $\rho\colon A' \to A' \otimes O(G) = A'[t_1,\ldots, t_r]$ be the coaction of $O(G)$ on $A'$. Then, for each $a' \in A'$, one may write

$$\rho(a') = \sum_{0\le\nu_1<p} \ldots \sum_{0\le\nu_r<p} (D_1^{\nu(1)} \ldots D_r^{\nu(r)}(a')/\nu_1!\ldots\nu_r!)t_1^{\nu(1)} \ldots t_r^{\nu(r)}$$

where $\nu(i) = \nu_i$ for each $1 \le i \le r$ ranges over the integers $0,1,\ldots, p-1$ and $D_i$ for each $1 \le i \le r$ stands for the derivation on $A'$ caused by the action of the i-th factor $\alpha_p$ of $G = \alpha_p \times \ldots \times \alpha_p$ (see DG-III, §2, 6.4; cf. 7.1.2 below). It is rather obvious that $D_i^p = 0$ and $D_iD_j = D_jD_i$ for all $1 \leq i, j \leq r$. Conversely, given a mutually commuting family $\{D_i: 1 \leq i \leq r\}$ of derivations with $D_i^p = 0$ for every i, one can define a coaction $\rho$ of $O(G)$ on $A'$ by means of the formula above, and hence obtains an action of $G = \alpha_p \times \ldots \times \alpha_p$ on Spec $A'$. Let us observe that the action of G is free if and only if there are elements $u_1', \ldots, u_r'$ of $A'$ such that $D_i(u_j') = \delta_{ij}$ (Kronecker's delta) for all $1 \leq i, j \leq r$. This follows, for instance, from 7.9.1 in the next section. Now the above formula for $\rho$ shows that for any $z \in K'^{\times}$ we have $\chi(z) \in A' \otimes O(G)$ if and only if $D_i(z)/z \in A'$ for all $1 \leq i \leq r$; it also shows that $\chi(z) = 1$ if and only if $D_i(z)/z = 0$ for all $1 \leq i \leq r$. This fact permits us to identify $L = \chi(K'^{\times}) \cap (A' \otimes O(G))$ with the (coordinate-wise additive) group of logarithmic derivatives

$$\{(D_1(z)/z, \ldots, D_r(z)/z): z \in K'^{\times}, D_i(z)/z \in A' \text{ for all } i\}$$

and $L_0 = \chi(A'^{\times})$ with the subgroup of all $(D_1(z)/z, \ldots, D_r(z)/z)$ such that $z \in A'^{\times}$. If in particular $r = 1$, we recover Samuel's L and $L_0$ as in 6.3.

6.4.2. LEMMA (Generalized Samuel's Lemma (6.3.1)). With the notations and assumptions as in 6.4, there is an exact sequence

$$0 \to L/L_0 \xrightarrow{i} C(A) \xrightarrow{j} C(A'),$$

where the map $i$ sends $\chi(z)$ (modulo $L_0$) to $\sum(v_{P'}(z)/e_{P'})P$ and $P = P' \cap A$ as in 6.3.1.

Proof. Our proof is essentially the same as that of Samuel's Theorem 6.3.1. Write an element of $C(A)$ in the form $\sum n_P P$. Then $j(\sum n_P P) = 0 \Longleftrightarrow \sum n_P e_{P'} P' = (a')$, a principal divisor with $a' \in K'^{\times}$. Thus $e_{P'} | v_{P'}(a')$ for all $P'$. Therefore, for any $P'$, there exists an $a \in K^{\times}$ such that $v_{P'}(a') = v_P(a)$ with $P' \cap A = P$, i.e., $a' = au'$, $u'$ being an invertible element of $A'_{P'}$. Thus $\chi(a') = \chi(a)\chi(u') = \chi(u')$. Since $\rho(A'_{P'}) \subseteq A'_{P'} \otimes O(G)$, it follows that $\chi(u') \in A'_{P'} \otimes O(G)$. Since $P'$ is arbitrary and $A'$ is a Krull domain, it follows that $\chi(a') \in A' \otimes O(G)$, i.e., $\chi(a') \in L$.

Moreover, $\chi(a') \in L_0 \Longleftrightarrow \chi(a') = \chi(w')$ with $w' \in A'^{\times}$ $\Longleftrightarrow \chi(a'/w') = 1$, i.e., $a = a'/w' \in K^{\times} \Longleftrightarrow (a')_{A'} = (a)_A$, i.e., $\sum n_P P$ is a principal divisor. Thus we have an injective map $\phi$: $\mathrm{Ker}(j) \longrightarrow L/L_0$.

We shall next show that $\phi$ is surjective. Let $\chi(z)$ be an element of L. Let P' be a prime ideal of height 1 of A'. It suffices to see that the ramification index $e_{P'}$ divides $v_{P'}(z)$. Notice that the induced coaction $\rho: A'_{P'} \longrightarrow A'_{P'} \otimes O(G)$ is free and that $A_P = (A'_{P'})^G$. Since $A'_{P'}$ is a discrete valuation ring, the assertion will follow from the following:

6.4.3. SUBLEMMA. _With the same notations and assumptions as in_ 6.4.2, _assume further that_ A' _is a discrete valuation ring and let_ $P' = tA'$ _be its unique maximal ideal with_ t _a uniformizing parameter. If_ $z \in K'^{\times}$ _is an element such that_ $\chi(z) \in A' \otimes O(G)$, _then_ $e_{P'}$ _divides_ $v_{P'}(z)$.

_Proof_. First assume that G is of height 1. Then $e_{P'} | p$. Hence $e_{P'} = 1$ or p. Let $z \in K'^{\times}$ be such that $\chi(z) \in A' \otimes O(G)$ and put $n = v_{P'}(z)$. We can assume that $p \nmid n$. Write $z = ut^n$ with $u \in A'^{\times}$. Then $\chi(t)^n = \chi(z)\chi(u^{-1}) \in A' \otimes O(G)$. Since L is an abelian group of exponent p, we have $\chi(t) \in A' \otimes O(G)$. Since $P' = tA'$, it follows that $\rho(P') \subseteq P' \otimes O(G)$. The induced coaction $\rho: A'/P' \longrightarrow A'/P' \otimes O(G)$ is of course free. Since $A/P \subseteq (A'/P')^G$, where $P = P' \cap A$, we have the inequality

$$\mathrm{rk}(G) = [K':K] \geqq [A'/P':A/P]e_{P'} \geqq [A'/P':(A'/P')^G]e_{P'}$$

$$= \mathrm{rk}(G)e_{P'}.$$

It follows that $e_{P'} = 1$ divides $v_{P'}(z) = n$.

In general we prove the assertion by the induction on the height of G. Let H be the kernel of the Frobenius map $F: G \longrightarrow G^{(p)}$. Then H is of height 1 and the height of G/H is smaller than N. Put $A'' = A'^H$ and $P'' = P' \cap A''$. The induced coaction $\rho: A'' \longrightarrow A'' \otimes O(G/H)$ is easily seen to be free. Let $e_{P'|P}$, $e_{P'|P''}$ and $e_{P''|P}$ denote the ramification indices of $P'|P$, $P'|P''$ and $P''|P$ respectively. Then $e_{P'|P} = e_{P'|P''}e_{P''|P}$. Suppose that $z \in K'^{\times}$ is such that $\chi(z) \in A' \otimes O(G)$. Since H is of height 1, we have $e_{P'|P''} \mid v_{P'}(z)$. Hence we can write $z = ux$ with $u \in A'^{\times}$ and $x \in K''^{\times}$, where K" denote the quotient field of A". Since then

$$\chi(x) = \chi(z/u) \in (A' \otimes O(G)) \cap (K'' \otimes O(G/H)) = A'' \otimes O(G/H),$$

it follows from the induction hypothesis that $e_{P''|P'} \mid v_{P''}(x)$. Therefore $e_{P'|P} = e_{P'|P''}e_{P''|P}$ divides $e_{P'|P''}v_{P''}(x) = v_{P'}(x) = v_{P'}(z)$, q.e.d.

A necessary and sufficient condition for a purely inseparable form of $\mathbb{A}^1$ or $\mathbb{S}^1$ to be trivial is given in the following:

6.5. THEOREM. Let k' be an arbitrary extension of k. Let A be a finitely generated k-algebra and let $A' = k' \otimes A$. Then we have:

(i) Assume that there is a finite and connected k-group scheme G which acts freely on k' in such a way that $k = k'^G$. Assume further that A' is k'-isomorphic to a polynomial ring k'[t] over k' in one variable t. Then A is k-isomorphic to a polynomial ring k[u] over k in one variable u if and only if A is a unique factorization domain and Spec(A) has a k-rational point.

(ii) Assume that A' is k'-isomorphic to $k'[t,t^{-1}]$ with

*an indeterminate* t. *Then* A *is* k-*isomorphic to* $k[u,u^{-1}]$ *with an indeterminate* u *if and only if the canonical homomorphism* $k'^{\times}/k^{\times} \longrightarrow A'^{\times}/A^{\times}$ *is an isomorphism.*

*Remark.* Suppose that k' is a finite purely inseparable and modular extension of k, so that there are elements $a_i \in k'$, $1 \leq i \leq r$, such that $k' \simeq k[a_1] \otimes \ldots \otimes k[a_r]$. Then the condition of (i) holds. In fact put $G = \alpha_{p(\nu(1))} \times \ldots \times \alpha_{p(\nu(r))}$, where $p(\nu(i)) = p^{\nu(i)} = [k[a_i]:k]$. Then G acts naturally and freely on k' in such a way that $k = k'^G$.

*Proof*. (i) Enough to prove the "if" part. By assumption there is a free coaction $\rho$: $k' \longrightarrow k' \otimes O(G)$ such that $k = k'^G$. This can be uniquely extended to a free coaction $\bar{\rho}$: $A' \longrightarrow A' \otimes O(G)$ such that $A = A'^G$. By virtue of Lemma 6.4.2 we have an exact sequence

$$0 \longrightarrow L/L_0 \longrightarrow C(A) \longrightarrow C(A').$$

Since $C(A) = 0$ by hypothesis, $L = L_0$. Moreover there is a k-homomorphism $\phi$: $A \longrightarrow k$. The extended k'-homomorphism $\bar{\phi}$: $A' \longrightarrow k'$ clearly commutes with the G-coactions. We may assume that $\bar{\phi}(t) = 0$. Then $\mathrm{Ker}(\bar{\phi}) = tA'$ is G-invariant, that is, $\bar{\rho}(tA') \subseteq tA' \otimes O(G)$. Hence $\chi(t) = (t \otimes 1)^{-1}\rho(t) \in A' \otimes O(G)$. Since $L = L_0$, there is an element $c \in k'^{\times}$ such that $\chi(t) = \chi(c)$. Here note that $A'^{\times} = k'^{\times}$. Replacing t by t/c we can assume that $\chi(t) = 1$, *i.e.*, $t \in A$. Since $A \supseteq k[t]$ and $k' \otimes A = k' \otimes k[t]$, it follows that $A = k[t]$.

(ii) The "only if" part: We may take u for t. It is then obvious that $k'^{\times}/k^{\times} = A'^{\times}/A^{\times}$.

The "if" part: Since $t \in A'^{\times}$ and $k'^{\times}/k^{\times} = A'^{\times}/A^{\times}$, there is $\lambda' \in k'^{\times}$ such that $u = \lambda'^{-1}t \in A^{\times}$. Then $A \supseteq k[u,u^{-1}]$ and $k' \otimes A = k' \otimes k[u,u^{-1}]$. This means immediately that we have $A = k[u,u^{-1}]$.

6.6. Remarks. (a) The first statement of Theorem 6.5 becomes false if any one of the two conditions is dropped. We shall give two examples.

(i) Let $\mathbb{F}_p$ be the prime field of characteristic p and let k be a purely transcendental extension $\mathbb{F}_p(t,u)$

of $\mathbb{F}_p$ with variables $t$ and $u$. Let $A = k[X,Y]/(Y^p = t + X + uX^p)$. Then, $A$ is a unique factorization domain, $\mathrm{Spec}(A)$ has no $k$-rational point and $k(t^{1/p},u^{1/p}) \otimes A$ is a polynomial ring over $k(t^{1/p},u^{1/p})$ in one variable. It is obvious that $A$ is not a polynomial ring over $k$ in one variable. Let us point out another curious fact about this example. If $p = 2$, let $k' = k(t^{1/2}, u^{1/2}) \supset k'' = k(t^{1/2}) \supset k$. Then, $k' \otimes A$ and $A$ are both unique factorization domains, but $k'' \otimes A$ is not, because $\mathrm{Pic}(k'' \otimes A) \simeq \mathbb{Z}/2\mathbb{Z}$ (see 6.11.2 below).

(ii) Let $A = k[X,Y]/(Y^p = X + \alpha X^p)$ with $\alpha \in k, \notin k^p$. $G = \mathrm{Spec}(A)$ is a one-dimensional unipotent group of Russell type which is not isomorphic to $G_a$. Hence $A$ is not a unique factorization domain, although it has a $k$-rational point $(0,0)$.

(b) Let $G = \mathrm{Spec}(B)$ be a one-dimensional unipotent group of Russell type, where $B = k[X,Y]/(Y^{p^n} = a_0X + \cdots + a_rX^{p^r})$ with $a_i$'s $\in k$ and some $a_j \notin k^p$ for $j \geq 1$. The point $(x = 0, y = 0)$ is a $k$-rational point of $G$ and $G-(x = 0, y = 0) = \mathrm{Spec}(A)$ with $A = B[1/x]$. Then $\mathrm{Spec}(A)$ is a purely inseparable form of $\mathbb{S}^1$. Let $k' = k(a_0^{p^{-n}}, \ldots, a_r^{p^{-n}})$ and let $A' = k' \otimes A$. It is easy to show that the cokernel of the canonical homomorphism $k'^{\times}/k^{\times} \to A'^{\times}/A^{\times}$ is $\mathbb{Z}/p^n\mathbb{Z}$.

Hence, 6.5 - (ii) shows that Spec(A) is a non-trivial form of $\mathbb{S}^1$.

6.7. We shall now take a closer look at the k-forms of $\mathbb{A}^1$. Let X be a k-form of $\mathbb{A}^1$ and let C be a k-normal completion of X, i.e., a complete k-normal curve defined over k containing X as a dense open set. Such a curve C exists and is determined by X up to k-isomorphisms. C-X is a one place point which might be singular and, for any field extension k' of k, $C_{k'}-X_{k'}$ is a one place point, too. Let $P_\infty$ = C-X. Then we have

6.7.1. LEMMA. With the notations as above, $P_\infty$ is rational over a purely inseparable extension of k.

Proof. Let k' be a perfect closure of k. Then $X_{k'}$ is k'-isomorphic to $\mathbb{A}^1_{k'}$. (Use the additive and multiplicative Theorem 90 of Hilbert.) Let C' be a k'-normalization of $C_{k'}$. Then C' is k'-isomorphic to $\mathbb{P}^1$. Hence $C'-X_{k'}$ is k'-rational. Since $P_\infty$ is dominated by the k'-rational point $C'-X_{k'}$, $P_\infty$ is rational over a purely inseparable extension of k, q.e.d.

By virtue of the well-known existence theorem of the Picard scheme for a proper k-scheme (see FGA, 236-02), we may consider the Picard scheme $\underline{\mathrm{Pic}}_{C/k}$ of the curve C, which is locally of finite type over k. The neutral component

$\underline{\mathrm{Pic}}^{o}_{C/k}$ of $\underline{\mathrm{Pic}}_{C/k}$ is of finite type over $k$. Some of the properties of the Picard scheme which we shall make use of later is summarized in

6.7.2. LEMMA. *Let $X$ be a proper scheme over $k$ such that $H^0(X, \mathcal{O}_X) = k$ and that $X$ has a k-rational point $P$. Then:*

(i) *For any $S \in \boxed{Sch}_k$, we have $\mathrm{Pic}(X_S) \cong \underline{\mathrm{Pic}}_{X/k}(S) \times \mathrm{Pic}(S)$ (direct product). $\underline{\mathrm{Pic}}_{X/k}(S)$ consists of the isomorphism classes of invertible sheaves $\mathcal{L}$ on $X_S$ whose restrictions on $P \times S$ are trivial. In particular, $\mathrm{Pic}(X_{k'}) \cong \underline{\mathrm{Pic}}_{X/k}(k')$.*

(ii) *Let $k'$ be a field extension of $k$. Then, $\underline{\mathrm{Pic}}_{X_{k'}/k'} \cong \underline{\mathrm{Pic}}_{X/k} \otimes k'$ and $\mathrm{Pic}^{o}_{X_{k'}/k'} \cong \underline{\mathrm{Pic}}^{o}{}_{X/k} \otimes k'$.*

*Proof.* (i) Let $f: X \to \mathrm{Spec}(k)$ be the structure morphism of a k-scheme $X$. Then the Picard scheme $\underline{\mathrm{Pic}}_{X/k}$ is identified with the first derived cohomology $R^1 f_*(G_{m,X})$, considered in the sense of (f.p.q.c)-cohomology. Moreover, the spectral sequence

$$E_2^{p,q} = H^p(S, R^q f_*(G_{m,X_S})) \Longrightarrow H^{p+q}(X_S, G_{m,X_S})$$

gives rise to an exact sequence,

$$0 \to H^1(S,G_{m,S}) \xrightarrow{f_S^*} H^1(X_S,G_{m,X}) \to H^0(S,R^1f_*(G_{m,X_S}))$$

$$\to H^2(S,G_{m,S}) \xrightarrow{f_S^*} H^2(X_S,G_{m,X_S}).$$

On the other hand, $f$ has a section $\sigma$, $f\sigma = 1$, such that $\sigma$(the only point of Spec(k)) = P. Hence $f_S\colon X_S \to S$ has a section $\sigma_S$. Therefore $H^2(S,G_{m,S}) \xrightarrow{f_S^*} H^2(X_S,G_{m,X_S})$ is injective and

$$0 \to H^1(S,G_{m,S}) \underset{\sigma_S^*}{\overset{f_S^*}{\rightleftarrows}} H^1(X_S,G_{m,X_S}) \to H^0(S,R^1f_*(G_{m,X_S})) \to 0$$

is split-exact. Thence follows that $\mathrm{Pic}(X_S) \cong \underline{\mathrm{Pic}}_{X/k}(S) \times \mathrm{Pic}(S)$ (cf. FGA-232).

(ii) Let $S' \in \boxed{\mathrm{Sch}}_{k'}$ and consider $S'$ as an object of $\boxed{\mathrm{Sch}}_k$. By the isomorphism of (i), $\underline{\mathrm{Pic}}_{X/k}(S') \cong \mathrm{Pic}(X\times S')/\mathrm{Pic}(S')$ $\cong \underline{\mathrm{Pic}}_{X_{k'}/k'}(S')$. Hence $\underline{\mathrm{Pic}}_{X_{k'}/k'} \cong \underline{\mathrm{Pic}}_{X/k} \otimes k'$. The isomorphism $\underline{\mathrm{Pic}}^o_{X/k} \otimes k' \cong \underline{\mathrm{Pic}}^o_{X_{k'}/k'}$ follows from the fact that the neutral component of a group scheme defined over a field is preserved by a base field extension.

6.7.3. LEMMA (M. Artin). Let $X$ be a k-regular proper integral scheme such that either $X$ has a k-rational point or $X$ is generically separable over $k$, and let $V$ be a smooth k-scheme. Then any rational map $f\colon V \to \underline{\mathrm{Pic}}_{X/k}$ is defined everywhere on $V$.

Proof. Let $k_s$ be the separable closure of $k$. Then $f$ is defined everywhere on $V$ if and only if $f \otimes k_s$: $V \otimes k_s \to \underline{\mathrm{Pic}}_{X/k} \otimes k_s$ is defined everywhere on $V \otimes k_s$. Hence, without any loss of generality, we may assume that $k$ is separably closed. Then in both cases, $X$ has a $k$-rational point $P$.

Let $U$ be a dense open set of $V$ on which $f$ is defined and let $g = f|_U : U \to \underline{\mathrm{Pic}}_{X/k}$. By virtue of 6.7.2 - (i), $g \in \underline{\mathrm{Pic}}_{X/k}(U)$ is representable by an invertible sheaf $\mathcal{L}$ on $X \times U$ such that $\mathcal{L}$ is trivial on $P \times U$. Note here that $X \times U$ is regular, hence locally factorial. Therefore there is a Weil divisor $D$ on $X \times U$ such that $\mathcal{L} = \mathcal{O}(D)$. Let $\bar{D}$ be the closure of $D$ in $X \times V$, which is regular too. Then, there is an invertible sheaf $\bar{\mathcal{L}}$ on $X \times V$ such that $\bar{\mathcal{L}} = \mathcal{O}(\bar{D})$ and $\mathcal{L} = \bar{\mathcal{L}}|_{X \times U}$. Then $\bar{\mathcal{L}}$ defines a $k$-morphism $\bar{f}: V \to \underline{\mathrm{Pic}}_{X/k}$ such that $\bar{f}|_U = g$. Hence $f = \bar{f}$. Thus $f$ is defined everywhere on $V$, q.e.d.

Going back to the notations of 6.7, we let $C$ be a $k$-normal completion of a $k$-form $X$ of $\mathbb{A}^1$ and let $P_\infty = C - X$. Let us suppose that $X = C - \{P_\infty\}$ has a $k$-rational point $P_0$.

6.7.4. LEMMA. $\underline{\mathrm{Pic}}^o_{C/k}$ is a $k$-smooth affine $k$-group scheme and there is a k-morphism $i: C-\{P_\infty\} \to \underline{\mathrm{Pic}}^{o}_{C/k}$ such that for any field extension $k'$ of $k$ and for $Q \in C(k')-\{P_\infty\}$, $i(Q) = Q-P_0$.

Proof. $\underline{\mathrm{Pic}}^{o}_{C/k}$ for k-scheme $C$ is smooth if $H^2(C, \mathcal{O}_C) = 0$ (see FGA, 236-15), which is certainly the case here because $C$ is a curve. Hence $\underline{\mathrm{Pic}}^{o}_{C/k}$ is k-smooth. On the other hand, $\underline{\mathrm{Pic}}^{o}_{C/k}$ is affine by virtue of Lemma 6.7.2 - (ii), since $C$ is rational over a purely inseparable extension of k. (The abelian rank of $\underline{\mathrm{Pic}}^{o}_{C/k}$ is zero.) Therefore, $\underline{\mathrm{Pic}}^{o}_{C/k}$ is a k-smooth affine connected algebraic k-group scheme.

We shall prove the second statement. Let $U = C - \{P_\infty\}$. $U$ is a smooth affine k-scheme of dimension 1. Since $U\times U$ is regular, the divisor $\Delta - P_0\times U$ on $U\times U$ is representable by an invertible sheaf $\mathcal{L}$ on $C\times U$ such that $\mathcal{L}$ is trivial on $P_0\times U$, where $\Delta$ is the diagonal of $U\times U$. Then $\mathcal{L}$ defines a k-morphism $i': U \to \underline{\mathrm{Pic}}_{C/k}$ such that $i'(P_0)$ = the neutral point of $\underline{\mathrm{Pic}}_{C/k}$. Since $U$ is connected, $i'$ is factored as follows: $i': U \xrightarrow{i} \underline{\mathrm{Pic}}^{o}_{C/k} \hookrightarrow \underline{\mathrm{Pic}}_{C/k}$. For any field extension $k'$ of $k$ and for $Q \in C(k') - \{P_\infty\}$, $i(Q) = Q-P_0$ since $\{\Delta - P_0\times U\} \cap (C\times Q) = (Q-P_0)\times Q$.

6.7.5. Before going to the next step, we shall recall the Riemann-Roch Theorem on a k-normal (not necessarily k-smooth) complete curve $C$ as given in 6.7. The notations being the same as at the beginning of 6.7, let D be a divisor on C whose support does not contain $P_\infty$. The degree of D, deg(D), is defined as follows: Let $\bar{k}$ be the algebraic closure of k and let $\phi: C_{\bar{k}} = C \otimes \bar{k} \to C$ be the canonical projection. The curve $C_{\bar{k}}$ is smooth outside of $\phi^{-1}(P_\infty)$. Then $\deg(D) := \deg_{C_{\bar{k}}}(\phi^{-1}(D))$. Then, the base change theorem (EGA, chap. III, (1.4.15)) and the Riemann-Roch Theorem on a (not necessarily smooth) complete curve over algebraically closed fields (cf. [14; chap. IV]) together give:

6.7.6. LEMMA. With the notations and assumptions* as above, we have

$$\dim H^0(C, \mathcal{O}(D)) - \dim H^1(C, \mathcal{O}(D)) = \deg(D) + 1 - \pi,$$

$\pi = \dim H^1(C, \mathcal{O}_C)$ = the arithmetic genus of C, and

$$\dim H^1(C, \mathcal{O}(D)) = \dim H^0(C, \omega \otimes \mathcal{O}(-D)),$$

where $\omega$ is the dualizing sheaf on C, which satisfies:

(a) $\omega \otimes \bar{k}$ is the dualizing sheaf on $C_{\bar{k}}$, which can be explicitly described (see [14; p.78]).

(b) $\omega$ is an invertible sheaf if and only if the local

---

* If $\pi = 0$, the support of D is allowed to contain $P_\infty$ (cf. Theorem 6.7.9).

ring of $C_{\bar{K}}$ at $C_{\bar{K}} - X_{\bar{K}}$ is a Gorenstein ring.

6.7.7. LEMMA. Let $C$ be a smooth complete k-curve such that $C' = C \otimes k'$ is k'-isomorphic to $\mathbb{P}^1_{k'}$, for a purely inseparable algebraic extension $k'$ of $k$. Assume that $C$ has a k-rational point if the characteristic $p$ is equal to 2. Then $C$ is k-isomorphic to $\mathbb{P}^1_k$.

Proof (M. Artin). We may assume that $k'$ is a simple extension of exponent 1, i.e., $[k':k] = p$. First of all, we shall show that G has a k-rational point if $p \neq 2$. Indeed, since $C'$ is k'-isomorphic to $\mathbb{P}^1_{k'}$, $C'$ has a k'-rational point $P'$. Then the cycle $pP'$ is k-rational on $C$. Let $K$ be the canonical divisor on $C$. Then $\deg(pP') = p$ and $\deg(K) = -2$. Let $n$ be a positive integer such that $p = 2n+1$ and consider the divisor $pP'+nK$ on $C$. Then $\deg(pP'+nK) = 1$. By virtue of 6.7.6 applied to $D = pP'+nK$, we have $\dim H^0(C, \mathcal{O}(pP'+nK)) = 2$ and $\dim H^1(C, \mathcal{O}(pP'+nK)) = 0$. Note here that the arithmetic genus of $C$ is zero. Thus there is a k-rational positive divisor $Q$ on $C$ such that $Q$ is linearly equivalent to $pP'+nK$. Since $\deg(pP'+nK) = 1$, $Q$ is a k-rational point of $C$ which does not ramify in $C'$.

In both cases ($p=2$ or $\neq 2$), $C$ has a k-rational point $Q$. Then the k-rational map $f: C \to \mathbb{P}^1_k$ defined by a complete linear system $|Q|$ is a k-isomorphism, since $f_{k'}: C' \to \mathbb{P}^1_{k'}$

is such.

6.7.8. Remark. If $p = 2$, Lemma 6.7.6 is false unless the existence of a k-rational point on $C$ is assumed, as shows the following example: Let $k = \mathbb{F}_2(t,u)$ be a purely transcendental extension of the prime field $\mathbb{F}_2$ with variables $t$ and $u$. Let $A = k[X,Y]/(Y^2 = t+X+uX^2)$. Then the completion $C$ of $\mathrm{Spec}(A)$ as imbedded canonically in $\mathbb{P}_k^2$ is a hypersurface defined by $Y^2 = tZ^2+XZ+uX^2$. Hence $C$ is a smooth complete k-curve such that $C_{k'}$ is k'-rational for $k' = k(t^{1/2}, u^{1/2})$. However $C$ is not k-rational. In fact, if $C$ were k-rational, $\mathrm{Spec}(A)$ should have sufficiently many k-rational points, whereas it is easy to see that $\mathrm{Spec}(A)$ has no k-rational point.

The foregoing lemmas 6.7.4 through 6.7.7 combined give the following:

6.7.9. THEOREM. *Let $C$ be a k-normal complete k-curve carrying a point $P_\infty$ such that $P_\infty$ is rational over a purely inseparable algebraic extension of $k$ and that $C - \{P_\infty\}$ is a k-form of $\mathbb{A}^1$. Assume that $C - \{P_\infty\}$ has a k-rational point $P_0$. Then, the following conditions are equivalent to each other:*

(i) *$i: C - \{P_\infty\} \to \mathrm{Pic}^o{}_{C/k}$ given in 6.7.4 is a closed immersion, and $\mathrm{Pic}^o_{C/k}$ is generated as a k-group scheme by* the image of i;

(ii) $\dim \mathrm{Pic}^{o}_{C/k} > 0$;

(iii) C is not k-isomorphic to $\mathbb{P}^1_k$, i.e., C is not k-rational;

(iv) $C-\{P_\infty\}$ cannot be embedded into a smooth complete k-curve.

Proof. (i) $\Longrightarrow$ (ii) and (iv) $\Longrightarrow$ (iii) are obvious. (ii) $\Longrightarrow$ (iii): Since $\dim \mathrm{Pic}^{o}_{C/k} = \dim H^1(C, \mathcal{O}_C)$ (cf. FGA,195-16), the arithmetic genus $\pi > 0$. Hence C is not k-isomorphic to $\mathbb{P}^1_k$. (iii) $\Longrightarrow$ (ii): It is well known that a k-normal complete curve with a k-rational point and with zero arithmetic genus is k-isomorphic to $\mathbb{P}^1_k$. Thence follows our assertion. (iii) $\Longrightarrow$ (iv): If $C-\{P_\infty\}$ can be imbedded into a smooth complete k-curve $\tilde{C}$, $\tilde{C}$ should be k-isomorphic to C. Lemma 6.7.6 then implies that C is k-isomorphic to $\mathbb{P}^1_k$. (ii) $\Longrightarrow$ (i): Let $\bar{k}$ be the algebraic closure of k, let $\bar{C} = C \otimes \bar{k}$ and let $\tilde{C}$ be the normalization of $\bar{C}$. Let $\tilde{P}_\infty$ be the point of $\tilde{C}$ over $P_\infty$. Let C' be a curve in $\mathbb{P}^2_{\bar{k}}$ defined as a hypersurface with the equation $Y^2Z = X^3$, and let $P'_\infty$ be the unique singular point of C'. We shall show that there is a $\bar{k}$-morphism $\phi: C' \to \bar{C}$. In fact, let $\tilde{\mathcal{O}}$, $\mathcal{O}'$ and $\bar{\mathcal{O}}$ be the local rings of points $\tilde{P}_\infty$, $P'_\infty$ and $P_\infty$ on $\tilde{C}$, C' and $\bar{C}$, respectively. With $\mathcal{O}'$ and $\bar{\mathcal{O}}$ identified with subrings of $\tilde{\mathcal{O}}$, we have only to show that $\mathcal{O}'$ dominates $\bar{\mathcal{O}}$. Identifying

$\tilde{\mathcal{O}}$ with the localization of a one-parameter polynomial ring $\bar{k}[t]$ at the ideal $(t)$, we then have $\mathcal{O}' = \bar{k} + t^2\tilde{\mathcal{O}}$. If $\mathcal{O}'$ does not dominate $\bar{\mathcal{O}}$, $\bar{\mathcal{O}}$ contains an element of $t\tilde{\mathcal{O}} - t^2\tilde{\mathcal{O}}$. Then, if $\hat{\mathcal{O}}$ and $\hat{\bar{\mathcal{O}}}$ denote the completions of $\tilde{\mathcal{O}}$ and $\bar{\mathcal{O}}$, respectively, we should have $\hat{\mathcal{O}} = \hat{\bar{\mathcal{O}}}$. Hence $\tilde{\mathcal{O}} = \bar{\mathcal{O}}$ because $\tilde{\mathcal{O}}$ is an $\bar{\mathcal{O}}$-module of finite type. This implies that $\dim \underline{\mathrm{Pic}}^{0}_{C/k} = 0$ (note (ii) $\Longleftrightarrow$ (iv)). Therefore $\mathcal{O}'$ dominates $\bar{\mathcal{O}}$. Now, the morphism $\phi\colon C' \to \bar{C}$ gives rise to a k-homomorphism $\bar{\rho}\colon \underline{\mathrm{Pic}}^{0}_{\bar{C}/\bar{k}} \to \mathrm{Pic}^{0}_{C'/\bar{k}}$. Let $\bar{i} = i \otimes \bar{k}\colon \bar{C} - \{P_\infty\} \to \mathrm{Pic}^{0}_{\bar{C}/\bar{k}}$. Then $\bar{\rho}\circ\bar{i}$ is given by $\bar{\rho}\circ\bar{i}\,(Q) = Q - P_0$ for $Q \in \bar{C}(\bar{k}) - \{P_\infty\}$. Since $\bar{C} - \{P_\infty\}$ and $C' - P'_\infty$ are isomorphic to $\mathbb{A}^1_{\bar{k}}$, it is easy to show that $\bar{\rho}\circ\bar{i}$ is a $\bar{k}$ isomorphism. In particular, $\bar{i}$ is a closed immersion. It follows that $i$ is a closed immersion, too. $\mathrm{Pic}^{0}_{\bar{C}/\bar{k}}$ is a generalized Jacobian variety $J_{\underline{m}}$ constructed from $\tilde{C} \simeq \mathbb{P}^1_{\bar{k}}$ with a module $\underline{m} = (\pi+1)\tilde{P}_\infty$, where $\pi = \dim H^1(C,\mathcal{O}_C)$, (cf. FGA, 195-16). Therefore, $\underline{\mathrm{Pic}}^{0}_{\bar{C}/\bar{k}}$ is generated as a $\bar{k}$-group scheme by the image of $\bar{i}$. Hence $\underline{\mathrm{Pic}}^{0}_{C/k}$ is generated as a k-group scheme by the image of i. Q.E.D.

The fact that $\underline{\mathrm{Pic}}^{0}_{\bar{C}/\bar{k}}$ is isomorphic to a generalized Jacobian variety and the results of [14; chap. V, Nos. 16, 17] imply

6.7.10. THEOREM. <u>The notations and assumptions being the same as in</u> 6.7.9, $\underline{\mathrm{Pic}}^{0}_{C/k}$ <u>is a commutative unipotent</u>

algebraic k-group. More precisely, $\mathrm{Pic}^{\circ}{}_{C/k}$ is a k-form of a product of Witt vector groups of finite lengths.

6.8. In the following two paragraphs, we shall give a complete classification of all k-forms of $\mathbb{A}^1$ whose arithmetic genera (defined as the arithmetic genera of their k-normal completions) are equal to 0 or 1, assuming only the existence of a k-rational point on each form. First of all, consider the case of arithmetic genus zero.

Let $a$ be an element of $k - k^p$ and let $n$ be a positive integer. Let $\phi: \mathbb{P}^1 \longrightarrow \mathbb{P}^{p^n}$ be the embedding of $\mathbb{P}^1$ into $\mathbb{P}^{p^n}$ given by $t \longmapsto (1, t, \cdots, t^{p^n - 1}, t^{p^n} - a)$, where $t$ is a parameter of $\mathbb{P}^1$. Let $P_\infty$ be a point of $\mathbb{P}^1$ defined by $t^{p^n} = a$. Denote by $X_{a,n}$ the image $\phi(\mathbb{P}^1 - \{P_\infty\})$. Then we have:

6.8.1. THEOREM. (i) Every k-rational k-form of $\mathbb{A}^1$ is k-isomorphic to either $\mathbb{A}^1$ or $X_{a,n}$ for suitable $a \in k - k^p$ and $n \in \mathbb{Z}^+$.*

(ii) $X_{a,n}$ is a k-rational k-form of $\mathbb{A}^1$ not k-isomorphic to $\mathbb{A}^1$.

(iii) $X_{a,n}$ is k-isomorphic to $X_{b,m}$ if and only if $m = n$ and there exist $\alpha, \beta, \gamma, \delta$ in $k^{p^n}$ such that $\alpha\delta - \beta\gamma \neq 0$

---

* If $p \geq 3$, all k-forms of $\mathbb{A}^1$ with arithmetic genus zero are k-rational (cf. 6.7.7).

and $(\alpha a + \beta)/(\gamma a + \delta) = b$.

Proof. (i) Let $X$ be any k-rational k-form of $\mathbb{A}^1$ and let $C$ be a k-normal completion of $X$. Theorem 6.7.9 and Lemma 6.7.1 then show that $C$ is k-isomorphic to $\mathbb{P}^1_k$ and $P_\infty = C - X$ is rational over a purely inseparable algebraic extension of $k$. Choose a parameter $t$ of $C$ $(\simeq \mathbb{P}^1_k)$ such that $t$ is finite at $P_\infty$. Suppose that $P_\infty$ is given by $t^{p^n} = a$ with $a \in k$, where $n$ is the smallest non-negative integer such that $P_\infty$ is given by $t^{p^n} = a$ with $a \in k$. If $n = 0$, $P_\infty$ is a k-rational point. Then $X$ is k-isomorphic to $\mathbb{A}^1$. Assume that $n > 0$. The divisor $P_\infty$ of $C$ has degree $p^n$. Hence, $\dim H^0(C, \mathcal{O}(P_\infty)) = p^n + 1$. Since $(1, 1/(t^{p^n} - a), t/(t^{p^n} - a), \dots, t^{p^n - 1}/(t^{p^n} - a))$ is a k-basis of the complete linear system $|P_\infty|$, the embedding $\phi: C \to P^{p^n}$ defined by $|P_\infty|$ is given by $t \longmapsto (1, t, \dots, t^{p^n - 1}, t^{p^n} - a)$. Then $X$ is k-isomorphic to $\phi(C - P_\infty)$ which is $X_{a,n}$.

(ii) Let $\phi: \mathbb{P}^1 \to \mathbb{P}^{p^n}$ be the embedding which defines $X_{a,n}$ and let $P_\infty$ be a point of $\mathbb{P}^1$ given by $t^{p^n} = a$. Then the point $\phi(P_\infty)$ is not k-rational, but rational over $k(a^{1/p^n})$. Therefore $X_{a,n}$ is a k-rational k-form of $\mathbb{A}^1$, not isomorphic to $\mathbb{A}^1$.

(iii) If $X_{a,n}$ is k-isomorphic to $X_{b,m}$, a k-isomorphism $\psi: X_{a,n} \to X_{b,m}$ extends to a k-isomorphism $\overline{\psi}$ between their k-normal completions $\overline{X}_{a,n}$ and $\overline{X}_{b,m}$, which sends the point $t^{p^n}$

$= a$ of $\overline{X}_{a,n}$ to the point $t'^{p^m} = b$ of $\overline{X}_{b,m}$. If $\overline{X}_{a,n}$ and $\overline{X}_{b,m}$ are identified with $\mathbb{P}^1_k$, $\overline{\psi}$ is given by $t \longmapsto (\alpha' t + \beta')/(\gamma' t + \delta')$ with $\alpha', \beta', \gamma', \delta' \in k$ such that $\alpha'\delta' - \beta'\gamma' \neq 0$. Then, setting $\alpha = \alpha'^{p^n}, \dots, \delta = \delta'^{p^n}$, we have $(\alpha a + \beta)/(\gamma a + \delta) = b$. (Clearly, we have $m = n$.) The "if" part is obvious.

6.8.2. We shall next consider the case where the arithmetic genus is equal to 1. It is known that a k-normal complete curve with the arithmetic genus 1 and with a k-rational point is k-isomorphic to a plane cubic defined by the hypersurface equation

$$Y^2Z + \lambda XYZ + \mu YZ^2 = X^3 + \alpha X^2Z + \beta XZ^2 + \gamma Z^3$$

with $\lambda, \mu, \alpha, \beta, \gamma \in k$, where we may assume that $\lambda = \mu = \alpha = 0$ if $p \neq 2,3$ and that $\lambda = \mu = 0$ if $p \neq 2$. Impose the condition that the above plane cubic have a one-place singular point, and we are led by direct calculations to the following two cases:

(1) $p = 3$ and $Y^2Z = X^3 + \gamma Z^3$ with $\gamma \notin k^3$,

(2) $p = 2$ and $Y^2Z = X^3 + \beta XZ^2 + \gamma Z^3$ with $\beta \notin k^2$ or $\gamma \notin k^2$.

Let $C$ be the plane cubic satisfying one of the above conditions and let $P_\infty$ be the singular point of $C$. Then

$P_\infty = (-\gamma^{1/3},0,1)$ in the first case and $P_\infty = (\beta^{1/2},\gamma^{1/2},1)$ in the second case. On the other hand, by virtue of 6.7.9, $C - \{P_\infty\}$ has a structure of a unipotent k-group, hence is k-isomorphic to a k-group of Russell type. The corresponding k-group of Russell type is:

(1)' $p = 3$ and $y^3 = x - \gamma x^3$ with $\gamma \notin k^3$,

(2)' $p = 2$ and $y^4 = x + \beta x^2 + \gamma^2 x^4$ with $\beta \notin k^2$ or $\gamma \notin k^2$.

It is easy to see that (1) corresponds to (1)'. However, it is not straightforward to see that (2) corresponds to (2)'. Let us therefore explain: Write the equation of (2)' in a homogeneous form

(2)" $T^4 + UV^3 + \beta U^2V^2 + \gamma^2U^4 = 0$

setting $y = T/V$ and $x = U/V$. The singular point of the curve (2)" is given by $(T,U,V) = (\gamma^{1/2},1,0)$. Let $t = T/U$ and $v = V/U$. Then, (2)" is $t^4 + v^3 + \beta v^2 + \gamma^2 = 0$. Hence $((t^2 + \gamma)/v)^2 = v + \beta$. Let $u = (t^2 + \gamma)/v$. Then we have $t^2 = u^3 + \beta u + \gamma$. This is the equation obtained from (2), setting $t = Y/Z$ and $u = X/Z$.

Summarizing the argument and computations above, we have

6.8.3. THEOREM.* <u>Every k-form of $\mathbb{A}^1$ with a k-rational point and with the arithmetic genus one is k-isomorphic to one of the following k-groups of Russell type</u>:

(1) $p = 3$ <u>and</u> $y^3 = x - \gamma x^3$ <u>with</u> $\gamma \notin k^3$,

*After the manuscript for the present paper was completed, the authors became aware of Clifford S. Queen's paper, "Non-conservative function fields of genus one, I", Arch. Math. (Basel), 22 (1971), 612-623. One can show that his Theorem 1 is equivalent to 6.8.3 here.

(2) $p = 2$ <u>and</u> $y^4 = x + \beta x^2 + \gamma^2 x^4$ <u>with</u> $\beta \notin k^2$ or $\gamma \notin k^2$.

6.8.4. <u>Remark</u>. According to Russell [11; Prop. 4.1], every k-form of $\mathbb{A}^1$ with arithmetic genus one is a principal homogeneous space for a k-group of Russell type. These k-groups are specifically given in our Theorem 6.8.3. Note that Russell misses the case (2) ($p = 2$) above in his paper (cf. <u>ibid</u>., p.539).

6.9. Let X be a k-form of $\mathbb{A}^1$ and let C be a k-normal completion of X. Since C is projective over k, the automorphism functor $S \in \boxed{\mathrm{Sch}}_k \longmapsto$ the group of all S-automorphisms of $C_S$, is representable by a k-scheme denoted as $\underline{\mathrm{Aut}}_{C/k}$, locally of finite type over k, whose neutral component, $\underline{\mathrm{Aut}}^{\circ}{}_{C/k}$, is a connected k-group scheme of finite type (cf. FGA, 221-10).

6.9.1 PROPOSITION (Rosenlicht [10], Russell [11]). <u>Let X and C be as above and let $P_\infty = C - X$. Assume that X has a k-rational point $P_0$. The following conditions are then equivalent to each other:</u>

(i) X <u>has a k-group structure with</u> $P_0$ <u>as the neutral point.</u>

(ii) X <u>is isomorphic to the underlying scheme of a k-group of Russell type.</u>

(iii) $\underline{Aut}_{C/k}(k_s)$ *is an infinite group, where* $k_s$ *is the separable closure of* k.

*If the arithmetic genus of* X *is non-zero, these conditions are equivalent to*

(iv) *There exists a surjective homomorphism* $\rho$ *of* k-*groups from* $\underline{Pic}^o{}_{C/k}$ *to a one-dimensional unipotent* k-*group* H *such that* $\rho \circ i$ *is an isomorphism with* $(\rho \circ i)(P_0)$ = *the neutral point of* H. (For the notations, see 6.7.4).

*Proof*. (i) ⟶ (ii): Any group structure on X is commutative, since dim X = 1. Hence it is a k-form of $G_a$. Then we are done by Russell [11, Theorem 2.1]. (ii) ⟹ (i): Obvious. (ii) ⟹ (iii): Let G be a k-group of Russell type isomorphic to X. Since translations by elements of G(R) give rise to R-automorphisms of $C_R$, G ↪ $\underline{Aut}_{C/k}$. Hence $\underline{Aut}_{C/k}(k_s)$ is an infinite group. (iii) ⟹ (ii): Russell [11; Theorem 4.2.].

Now assume that the arithmetic genus of X is non-zero and let H be a group structure given on X with $P_0$ as the neutral point. For any field k' over k, define $(\rho \circ i)(k')$: X(k') → H(k') by $P_0 \longmapsto$ the neutral point of H and $Q \mapsto Q - P_0$ (sum with respect to the group law of H). Then $\rho \circ i$ is a k-isomorphism. Since $\underline{Pic}^o{}_{C/k}$ is generated as a group scheme by the image of i, $\rho \circ i$ extends to a surjective homomorphism $\rho : \underline{Pic}^o_{C/k} \to H$. This implies

(i) $\Longrightarrow$ (iv). (iv) $\Longrightarrow$ (i): Obvious.

6.9.2. THEOREM. (Rosenlicht [10], Russell [11]). Let $G$ be a $k$-group of Russell type. Then the function field $k(G)$ over $k$ is rational if and only if $p = 2$ and $G$ is defined by an equation $y^2 = x + ax^2$ with $a \in k$, $a \notin k^2$.

Proof. The "if" part is straightforward. We shall prove the "only if" part. Since $k(G)$ is rational, $G$ has a $k$-rational point $P$ other than the neutral point. The translation by $P$ induces a $k$-automorphism of the underlying scheme of $G$, hence a $k$-normal completion $\mathbb{P}^1$ of $G$. This automorphism of $\mathbb{P}^1$ is given by $t \longmapsto (at + b)/(ct + d)$ with $a,b,c,d \in k$ and $ad - bc \neq 0$, and leaves the point $\mathbb{P}^1 - G$ fixed. Choose a parameter $t$ of $\mathbb{P}^1$ such that $t$ is finite at $\mathbb{P}^1 - G$. Let $t(\mathbb{P}^1 - G) = \alpha$. Then $(a\alpha + b)/(c\alpha + d) = \alpha$, or $c\alpha^2 + (d - a)\alpha - b = 0$. Therefore $[k(\alpha): k] \leq 2$.

On the other hand, $\alpha$ is purely inseparable over $k$ by 6.7.1. Hence $p = 2$ or $\alpha$ is $k$-rational. In the latter case, $G$ is isomorphic to $G_a$. (This case is excluded.) Hence $p = 2$, $d = a$ and $c\alpha^2 = b$. Since $c \neq 0$, $\alpha^2 = b/c$.

Set again $\alpha^2 = a$ with $a \in k$, $\notin k^2$. Let $P_\infty = \mathbb{P}^1 - G$. Then the dimension of the complete linear system $|P_\infty|$ is two and the rational map $\psi$ defined by $|P_\infty|$ is an embedding of $\mathbb{P}^1$ into $\mathbb{P}^2$ given by $t \longmapsto (1,t,t^2-a)$ (see 6.8.1). Then the image $\psi(\mathbb{P}^1 - P_\infty)$ is a curve in $\mathbb{A}^2$ defined by $y^2 = x + ax^2$. Q.E.D.

6.9.3. When our theorems 6.7.9, 6.8.1 and 6.8.3 are reviewed in conjunction with Rosenlicht-Russell's 6.9.1 and 6.9.2, it may be said that at least over a separably closed field $k = k_s$ all those k-forms of $\mathbb{A}^1_k$ have been completely determined which either possess infinitely many k-automorphisms or are of arithmetic genus $\leq 1$. Are there, then, k-forms of $\mathbb{A}^1_k$ of arithmetic genus $> 1$ possessing only finitely many k-automorphisms? The answer is in the affirmative, as seen from the following example communicated to us by M. Rosenlicht:

<u>Example</u> (M. Rosenlicht). Let $k = k_s$ and $p > 3$, and fix an element $a \in k - k^p$. Consider

$$C: \quad y^2 z^{p-2} = x^p - az^p \qquad (\text{in } \mathbb{P}^2_k),$$

a hyperelliptic k-curve of arithmetic genus $(p-1)/2$. Of the two singular points $P_1 = (a^{1/p}, 0, 1)$, $P_2 = (0, 1, 0)$ of $C$, $P_1$ is k-normal and $P_2$ is not. However, $P_2$ is dominated by only one place of the function field $k(C)$. Let $X := C - \{P_1\}$ whose equation on $\mathbb{A}^2_k$ is

$$X: \quad \zeta^{p-2} + a\zeta^p = \xi^p \qquad (\zeta = z/y,\ \xi = x/y).$$

At the unique singular point $P_2 = (\xi = 0, \zeta = 0)$ of $X$, the functions

$$t = \xi^q/\zeta^{q-1},\ u = \zeta/\xi \qquad \text{where } q = (p-1)/2$$

are of order 1 and 2, respectively, and may be taken as uniformizing parameters of $X$ at $(0,0)$. In fact, the correspondence $(\xi, \zeta) \mapsto (t, u)$

gives an anti-regular birational transformation of X to Y given in $\mathbb{A}_k^2$ by

$$Y: \quad t^2 - at^2u^p - u = 0$$

with the inverse formulae $\xi = tu^{q-1}$, $\zeta = tu^q$. As Y is k-smooth at $Q = (t = 0,\ u = 0)$ and $P_2$ is a one-place point, the local ring $\mathcal{O}_{Q,Y}$ is the integral closure of $\mathcal{O}_{P_2,X}$. We conclude that the k-normalization $\tilde{X}$ of X is a k-smooth affine curve with a unique k-normal singular point $P_1$ at infinity. Since moreover the geometric genus of $\tilde{X}$ is clearly zero, $\tilde{X}$ must be a k-form of $\mathbb{A}_k^1$, of arithmetic genus $q = (p-1)/2 > 1$. Let us now show that $\tilde{X}$ has only two k-automorphisms. Write the hyperelliptic field $k(\tilde{X}) = k(C)$ as

$$k(\tilde{X}) = k(v,w) \quad \text{with } v = x/z,\ w = y/z \quad \text{and } w^2 = v^p - a.$$

Then, k(v) is the unique subfield of index 2 generated over k by the ratios of the differentials of the first kind. It is therefore invariant under any k-automorphism of $k(\tilde{X})$. Let $\sigma$ be an arbitrary k-automorphism of $\tilde{X}$, which we extend uniquely to a k-automorphism of its k-normal completion $\tilde{X} \cup \{P_1\}$. It then causes a k-automorphism $\sigma^*$ of $k(\tilde{X}) = k(v,w)$ leaving the subfield k(v) invariant. Thus,

$$\sigma^*(v) = \frac{\alpha v + \beta}{\gamma v + \delta}, \quad \alpha\delta - \beta\gamma = 1$$

for some $\alpha, \beta, \gamma, \delta \in k$ holds. On the other hand, $\sigma P_1 = P_1$ by the extended action of $\sigma$, and $P_1 = (v = a^{1/p},\ w = 0)$, so that

$$\sigma P_1 = \sigma(a^{1/p}, 0) = (\sigma^*(a^{1/p}), \sigma^*(0)) = P_1 = (a^{1/p}, 0).$$

Thus, $(\alpha a^{1/p} + \beta)/(\gamma a^{1/p} + \delta) = a^{1/p}$ follows. Since $p \neq 2$, this implies $\gamma = \beta = 0$ and $\alpha = \delta \neq 0$. Therefore, $\sigma^*(v) = v$ and hence $\sigma^*(w^2) = \sigma^*(v^p - a) = w^2$. We have thus established that the two k-automorphisms induced by $(v,w) \mapsto (v, \pm w)$ are all and only k-automorphisms of $\tilde{X}$.

6.10. Let $G = \mathrm{Spec}(A)$ be a k-group of Russell type and let $A = k[X,Y]/(Y^{p^n} = X + a_1 X^p + \cdots + a_m X^{p^m})$ with $a_1, \cdots, a_m \in k$ and not all of $a_1, \cdots, a_m \in k^p$. Let $C$ be a k-normal completion of $\bar{G}$ and let $P_\infty = C - G$. We are interested in the structure of the divisor class group $C(A) = \mathrm{Pic}\,(G)$ of A.

6.10.1 THEOREM. *Let the notations be as above. We have then the following two exact sequences*

(i) $0 \to \mathbb{Z} \overset{j}{\to} \mathrm{Pic}(C) \to C(A) \to 0$

(ii) $0 \to \underline{\mathrm{Pic}}^o{}_{C/k}(k) \to C(A) \to \mathbb{Z}/p^r\mathbb{Z} \to 0$

where $p^r$ is the ramification index of the place corresponding to $P_\infty$ in $\bar{k}(G_{\bar{k}})$ and $p^r \leq p^n$.

*Proof*. The map $\rho$ is obtained by restricting invertible sheaves over $C$ down to $G$. $\rho$ is surjective. In fact, let $D$ be a divisor on $G$. Let $\bar{D}$ be the closure of $D$ on $C$. Since $C$ is k-normal, $\bar{D}$ is given by an invertible sheaf on $C$, whose restriction on $G$ is $\mathcal{O}(D)$. Next, if a divisor $E$ on $C$ is linearly equivalent to $0$ on $G$, there is a function $f \in k(G)$ and an integer $s$ such that $E - sP_\infty = (f)$, the divisor of $f$. Define the map $j$ assigning $1 \in \mathbb{Z}$ to $P_\infty$.

It is then easy to see that the sequence (i) is exact. On the other hand, we have an exact sequence

$$0 \to \underline{\mathrm{Pic}}^{\mathrm{o}}{}_{C/k}(k) \to \mathrm{Pic}(C) \overset{j'}{\to} \mathbb{Z} \to 0$$

where the neutral point of $G$ is a generator of $\mathbb{Z} \simeq \mathrm{Pic}(C)/\underline{\mathrm{Pic}}^{\mathrm{o}}{}_{C/k}(k)$ up to algebraic equivalence. $\underline{\mathrm{Pic}}^{\mathrm{o}}{}_{C/k}(k) \cap j(\mathbb{Z}) = (0)$, since the degree of every element of $\underline{\mathrm{Pic}}^{\mathrm{o}}{}_{C/k}(k)$ is zero. Moreover, $j'(P_\infty) = p^r$. From these observations, we get the second exact sequence.

6.11. We shall now carry out concrete and explicit calculations in some special cases in order to clarify the structure of the divisor class groups of Russell type k-groups.

6.11.1. LEMMA. *Let $k$ be a non-perfect, separably closed field of characteristic $p \neq 0$ and let $A = k[X,Y]/(Y^p = X - a_1X^p - \cdots - a_rX^{p^r})$, where $a_1, \cdots, a_r \in k$ and $\notin k^p$ such that $a_1, \cdots, a_r$ are p-independent over $k^p$. Then the divisor class group $C(A)$ of $A$ is a $\mathbb{Z}/p\mathbb{Z}$-module with generators indexed by elements of $G(k)$, where $G = \mathrm{Spec}(A)$. In particular, $C(A)$ is an infinite p-group, except when $p = 2$ and $r = 1$.*

*Proof*. The proof consists of several steps.

(I) Let $\lambda_i^p = a_i$ with $\lambda_i \in k^{1/p}$ for $1 \leq i \leq r$.

Let $k' = k(\lambda_1, \cdots, \lambda_r)$. Since $\lambda_1, \cdots, \lambda_r$ are p-independent over $k$, there is a set of k-derivations $(D_1, \cdots, D_r)$ in $k'$ such that $D_i(\lambda_j) = \delta_{ij}$ and $D_i D_j = D_j D_i$ for $1 \leq i,j \leq r$. Let $A' = k' \otimes A$. The derivations $D_1, \cdots, D_r$ can be uniquely extended on $A'$ and satisfy: $D_i \neq 0$, $D_i^p = 0$, $D_i D_j = D_j D_i$, $D_i(\lambda_j) = \delta_{ij}$ for $1 \leq i,j \leq r$ and $A = \{a' \in A' | D_i(a') = 0$ for $1 \leq i \leq r\}$. Moreover, writing $A = k[x,y]$, let $t = y + \lambda_1 x + \cdots + \lambda_r x^{p^{r-1}}$ ; then $x = t^p$ $y = t + \lambda_1 t^p + \cdots + \lambda_r t^{p^r}$ and $A' = k'[t]$. For $1 \leq i \leq r$, $D_i(t) = x^{p^{i-1}} = t^{p^i}$. Therefore we can apply 6.4.1 and 6.4.2 to $A'$, $A$ and the quotient field $k'$ of $A'$. Thus we have $L/L_0 \simeq C(A)$, noting that $C(A') = 0$.

(II) We shall calculate the group $L$. Let $z \in K'^{\times} = k'(t)^{\times}$. Write $z = f(t)/g(t)$ with $f(t), g(t) \in k'[t]$ such that $f(t)$ and $g(t)$ have no common divisors. Note here that a monic irreducible polynomial in $k'[t]$ should be of the form $t^{p^\ell} - d$ with $\ell \in \mathbb{N}$, $d \in k'$ and $d \notin k'^p$ if $\ell > 0$. Hence $z$ can be written in the form* $z = c \prod_{j=1}^{s} (t^{p(\ell_j)} - d_j)^{\alpha_j}$, where $c \in k'$, $\alpha_j \in \mathbb{Z}^+$ and $d_j \in k'$, $d_j$'s satisfying the condition

(C) $d_j \neq d_{j'}$ if $\ell_j = \ell_{j'}$, and $d_j \notin k'^p$ if $\ell_j > 0$.

---

* To avoid triple superfix we write $X^{p(\nu)}$ for the $p^\nu$-th power of the entity $X$.

Then for any $D_i$ $(1 \leq i \leq r)$, $D_i(z)/z = (D_i(c)/c) + \sum_{j=1}^{s}\alpha_j D_i(t^{p(\ell_j)} - d_j)/(t^{p(\ell_j)} - d_j)$. Meanwhile $D_i(t^{p^\ell} - d)/(t^{p^\ell} - d) = -D_i(d)/(t^{p^\ell} - d)$ if $\ell \neq 0$ and $D_i(t^{p^\ell} - d)/(t^{p^\ell} - d) = (t^{p^i} - D_i(d))/(t - d)$ if $\ell = 0$. Hence

$$D_i(z)/z = D_i(c)/c + \sum_{j'}\alpha_{j'}(t^{p^i} - d_{j'}^{p^i})/(t - d_{j'})$$
$$+ \sum_{j=1}^{s}\alpha_j(p^{\ell_j}d_j^{p^i} - D_i(d_j))/(t^{p(\ell_j)} - d_j),$$

where $\sum_{j'}$ stands for a partial sum in $\sum_j$ for $j'$ with $\ell_{j'} = 0$. Thus $D_i(z)/z \in k'[t]$ if and only if $\sum_{j=1}^{s}\alpha_j(p^{\ell_j}d_j^{p^i} - D_i(d_j))/(t^{p(\ell_j)} - d_j) = 0$. However, this is equivalent to

$$\sum_{j=1}^{s}\{\alpha_j(p^{\ell_j}d_j^{p^i} - D_i(d_j))\prod_{h\neq j}(t^{p(\ell_h)} - d_h)\} = 0.$$

Put $t = d_j^{p(-\ell_j)}$ in the last equation. Then

$$\alpha_j(p^{\ell_j}d_j^{p^i} - D_i(d_j))\prod_{h\neq j}(d_j^{p(\ell_h - \ell_j)} - d_h) = 0$$

where $\prod_{h\neq j}(d_j^{p(\ell_h - \ell_j)} - d_h) \neq 0$ by the above condition (C). Thus $p^{\ell_j}d_j^{p^i} = D_i(d_j)$ if $\alpha_j \not\equiv 0$ (modulo p). If $\alpha_j \not\equiv 0$ (modulo p), this implies that $d_j^{p^i} = D_i(d_j)$ if $\ell_j = 0$ and $D_i(d_j) = 0$ if $\ell_j \neq 0$. Therefore L is generated by

$(D_1(t - d)/(t - d),\ldots,D_r(t - d)/(t - d))$ with $D_1(d) = d^p$, $\ldots,D_r(d) = d^{p^r}$.

(III) Let $d$ be an element of $k'$ such that $D_i(d) = d^{p^i}$ for $1 \leq i \leq r$. It is easy to show that $d$ can be written in the form $d = c_0+c_1\lambda_1+\cdots+c_r\lambda_r$ with $c_0,\ldots,c_r \in k$ such that

$$c_1 = c_0^p+a_1c_1^p+\cdots+a_rc_r^p, \quad c_2 = c_0^{p^2}+a_1^pc_1^{p^2}+\cdots+a_r^pc_r^{p^2},$$

$$\ldots, \quad c_r = c_0^{p^r}+a_1^{p^{r-1}}c_1^{p^r}+\cdots+a_r^{p^{r-1}}c_r^{p^r}.$$

Then $c_2 = c_1^p,\ldots,c_r = c_1^{p^{r-1}}$. Hence $c_1 = c_0^p+a_1c_1^p+\cdots+a_rc_1^{p^r}$. Therefore $(y = c_0,\ x = c_1)$ satisfies the defining equation of $A$, i.e., $(y = c_0,\ x = c_1) \in G(k)$. Conversely if $(y = c_0,\ x = c_1) \in G(k)$, $d = c_0+c_1\lambda_1 + \cdots + c_1^{p^{r-1}}\lambda_r$ satisfies $D_i(d) = d^{p^i}$ for $1 \leq i \leq r$. Hence it gives rise to a generator $(D_1(t - d)/(t - d),\ \ldots,\ D_r(t - d)/(t - d))$ of $L$.

(IV) Now $D_i(t-d)/(t-d) = t^{p^i-1} + t^{p^i-2}d + \cdots + td^{p^i-2} + d^{p^i-1}$, where $t^{p^i-1} = D_i(t)/t$ and $d^{p^i-1} = D_i(d)/d$. On the other hand, we have a surjective homomorphism $\phi : L \to \mathfrak{F}$: = the group generated by the coefficients of $t^{p^r-2}$ in the last coordinate of elements of $L$. We call $\mathfrak{F}$ the last coordinate group of $L$. It is clear that $\phi(L_0) = 0$ and $\mathfrak{F} \simeq G(k)$. Therefore $C(A)$ is generated by $G(k)$ and $C(A)$ is an infinite p-group except when $p = 2$ and $r = 1$. When

$p = 2$ and $r = 1$, it is easy to show that $C(A) \simeq \mathbb{Z}/2\mathbb{Z}$.

6.11.2 Remarks. (1) If $a_1, \cdots, a_r$ are not p-independent over $k^p$, the computation of $C(A)$ becomes more complicated.

(2) Consider the case where $A = k[x,y]/(y^p = x - ax^p)$ wiht $a \in k$ and $a \notin k^p$. If $p = 2$, $C(A) = \mathbb{Z}/2\mathbb{Z}$. If $p = 3$, $0 \to G(k) \to C(A) \to \mathbb{Z}/3\mathbb{Z} \to 0$ is exact (cf. 6.10.1). If $p = 3$, a k-normal completion $C$ of $G = \mathrm{Spec}(A)$ is isomorphic to the plane cubic $Y^3 = XZ^2 - aX^3$ in $\mathbb{P}^2_k$, all points of which are inflection points. The addition of the group $G(k)$ which is induced from the usual addition in $G_a \times G_a$ coincides with the addition caused on $C(k) - (1, -a^{1/3}, 0)$ in the usual fashion to define the addition on a plane cubic. Here $(X = 0, Y = 0, Z = 1)$ is the neutral point. Moreover, $i: G \longrightarrow \underline{\mathrm{Pic}}^0_{C/k}$ is an isomorphism.

6.12. In 6.11, the calculation of $C(A)$ was made over a non-perfect separably closed field $k$. In order to calculate $C(A)$ over a not necessarily separably closed field, we need the following

6.12.1 LEMMA. *Let $k$ be a field of characteristic $p \neq 0$ and let $k'$ be a separable algebraic closure of $k$ with Galois group $\Gamma$. Let $A$ be a smooth k-algebra such that $A^\times = k^\times$ and let $A' = k' \otimes A$. Then $C(A) = C(A')^\Gamma$ = the group consisting of $\Gamma$-invariant elements of $C(A')$.*

Proof. Since $A'$ is a smooth $k'$-algebra, $C(A')$ is isomorphic to the group of isomorphism classes of projective $A'$-modules of rank 1. Let $P$ be a projective $A'$-module of rank 1 such that the corresponding class $\{P\}$ in $C(A')$ is $\Gamma$-invariant. For any $\sigma \in \Gamma$, denote by $\phi_\sigma$ the isomorphism $P \to {}^\sigma P = (k',\sigma) \underset{k'}{\otimes} P$. $\phi_\sigma$ is determined up to a scalor multiple in $k'^\times$ since $A'^\times = k'^\times$. For any $\sigma,\tau \in \Gamma$, $\phi_{\sigma\tau}$ and ${}^\sigma\phi_\tau \cdot \phi_\sigma$ differ by a scalor multiple $a(\sigma,\tau) \in k'^\times$, i.e., $\phi_{\sigma\tau} = a(\sigma,\tau)^\sigma\phi_\tau \cdot \phi_\sigma$. For $\sigma,\tau,\nu \in \Gamma$, we have ${}^\sigma a(\tau,\nu)a(\sigma,\tau\nu) = a(\sigma\tau,\nu)a(\sigma,\tau)$. Thus $a(-,-)$ is a 2-cocycle of $\Gamma$ with values in $k'^\times$. Since $H^1(\Gamma,k'^\times) = (0)$ (Hilbert's Theorem 90), there exists $b(\sigma)$ for all $\sigma \in \Gamma$ such that $b(\sigma) \in k'^\times$ and $a(\sigma,\tau) = {}^\sigma b(\tau)b(\sigma\tau)^{-1}b(\sigma)$. Let $\phi'_\sigma = b(\sigma)\phi_\sigma$. Then $\phi'_{\sigma\tau} = {}^\sigma\phi'_\tau \cdot \phi'_\sigma$. In other words, $\{\phi'_\sigma | \sigma \in \Gamma\}$ defines a descent data on $P$ for $k'/k$. Therefore there exists a projective $A$-module $\tilde{P}$ of rank 1 such that $P = k' \otimes \tilde{P}$. Assigning $P$ to $\tilde{P}$, we have an inverse map to the canonical homomorphism $C(A) \to C(A')$.

6.12.2 LEMMA. Let $A = k[X,Y]/(Y^p = X - a_1X^p - \cdots - a_rX^{p^r})$ with $a_1,\cdots,a_r \in k$ and $\notin k^p$ such that $a_1,\cdots,a_r$ are p-independent over $k^p$. Then the divisor class group $C(A)$ of $A$ is a $\mathbb{Z}/p\mathbb{Z}$-module which is isomorphic to $C(A_s)^\Gamma$, where $A_s = k_s \otimes A$, $k_s$ being a separable algebraic closure of $k$ with Galois group $\Gamma$. $C(A_s)$ is generated by $G(k_s)$, where $G = \mathrm{Spec}(A)$.

Proof. This follows immediately from Lemmas 6.11.1 and 6.12.1.

6.12.3 Remarks. (1) Let $p = 3$, let $k = \mathbb{F}_3(t)$ be a simple transcendental extension of the prime field $\mathbb{F}_3$ and let $A = k[X,Y]/(Y^3 = X-tX^3)$. Then $G(k) = \{(X = 0, Y = 0)\}$ with $G = \mathrm{Spec}(A)$. However if $k_s$ is an algebraic separable closure of $k$, $G(k_s)$ is an infinite group. This implies that $A[1/x]$ is a unique factorization domain, while $k_s \otimes A[1/x]$ is not a unique factorization domain.

(2) Let $A = k[x,y]/(y^p = x - ax^p)$ with $a \in k$ and $a \notin k^p$. Imbed $G = \mathrm{Spec}(A)$ canonically into $\mathbb{P}^2$ and take the completion $C$ of $G$. $C$ is defined by the equation $Y^p = XZ^{p-1} - aX^p$ in the homogeneous coordinates $(X,Y,Z)$ of $\mathbb{P}^2$ such that $x = X/Z$ and $y = Y/Z$. The point at infinity $(1, -a^{1/p}, 0)$ is not k-rational. Moreover $C$ is k-normal (but not geometrically normal except when $p = 2$), and $\underline{\mathrm{Pic}}^o{}_{C/k}$ is a smooth affine group of dimension $(p - 1)\cdot(p - 2)/2$. As indicated in 6.11.2, if $p = 3$, $i: G \to \underline{\mathrm{Pic}}^o{}_{C/k}$ is an isomorphism.

6.13. Finally we shall make one remark on $\mathrm{Ext}_{\mathrm{cent}}(G,G_m)$ for a one-dimensional unipotent group of Russell type $G$. If $k$ is an algebraically closed field of arbitrary characteristic and if $G$ is a connected algebraic linear k-group, $\mathrm{Ext}_{\mathrm{cent}}(G,G_m)$ is isomorphic to $\mathrm{Pic}(G)$. However this result no longer holds if $k$ is not algebraically closed. A counter-example is given by a one-dimensional unipotent group of Russell type. In fact, we shall see below that $\mathrm{Ext}_{\mathrm{cent}}(G,G_m) \neq C(A)$ $(= \mathrm{Pic}\, G)$ if $G = \mathrm{Spec}(A)$ with $A = k[x,y]/y^p = x - ax^p$, $a \in k$ and $a \notin k^p$, while $C(A) \neq 0$. We shall prove first a

6.13.1. LEMMA. <u>Let $G = \mathrm{Spec}(A)$ be a commutative unipotent k-group. Then the canonical map $\mathrm{Ext}_{\mathrm{cent}}(G,G_m) \to C(A) = H^1(G,G_m)$ is injective</u>

<u>Proof</u>. (I) First of all, note that all extensions of $G$ by $G_m$ are commutative. Indeed, let $E$ be an extension of $G$ by $G_m$. It suffices to show that $E$ is commutative when $k$ is algebraically closed. Then $E$ is k-solvable. Hence $E = G \times G_m$ (a direct product). In particular, $E$ is commutative.

(II) Let $E$ be an extension of $G$ by $G_m$. Let $E = \mathrm{Spec}(B)$. Since $G_m$ acts on $E$ on the left, $B = A[P,P^{-1}]$ = the symmetric tensor algebra over $A$ generated by $P$ and $P^{-1}$ with relations $P^m P^{-n} = P^{m-n}$ for $m, n \in \mathbb{Z}$, where $P$ is a projective $A$-module of rank 1. The canonical map $\mathrm{Ext}_{\mathrm{cent}}(G,G_m) \to C(A)$ sends $E$ to $P^{-1}$. If $P \simeq A$, then $E = G \times G_m$ as a scheme. The multiplication of $E$ is given by a 2-cocycle of $G$ with values in $G_m$; <u>i.e.</u>, for $x, x' \in G$ and $g, g' \in G_m$, $(g,x)(g',x') = (gg'f(x,x'),xx')$, where $f(x,x')$ is a 2-cocycle such that $f(x,e) = f(e,x) = 1$. Since

$(A \underset{k}{\otimes} A)^{\times} = k^{\times}$, $f(x, x')$ is a constant. However since $f(x, e) = f(e, x) = 1$, $f(x, x') = 1$ for all $x, x' \in G$. Therefore $(g, x)(g', x') = (gg', xx')$. Thus $E = G \times G_m$ as a group.

6.13.2 PROPOSITION. Let $k$ be a separably closed field of characteristic $p > 2$ and let $G = \mathrm{Spec}(A)$ be a one-dimensional unipotent $k$-group of Russell type with $A = k[x, y]/(y^p = x - ax^p)$, $a \in k$ and $a \notin k^p$. Then $\mathrm{Ext}_{\mathrm{cent}}(G, G_m) \neq C(A) = H^1(G, G_m)$.

Proof. Let $E$ be an extension of $G$ by $G_m$. The affine ring of $E$ is isomorphic to $A[P, P^{-1}]$, where $P$ is a projective $A$-module of rank 1. Let $\Delta : A \longrightarrow A \otimes A$, $q_1 : A \longrightarrow A \otimes A$ and $q_2 : A \longrightarrow A \otimes A$ be the comultiplication of $A$, $q_1(b) := b \otimes 1$ and $q_2(b) := 1 \otimes b$, respectively. Denote by $\Delta^*(P)$ (or $q_i^*(P)$) the $A \otimes A$-module $\Delta^*(P) := P \underset{A}{\otimes} (A \otimes A, \Delta)$ (or $q_i^*(P) := P \underset{A}{\otimes} (A \otimes A, q_i)$). Then $P$ satisfies the relation $\Delta^*(P) = q_1^*(P) \otimes q_2^*(P)$, where $\otimes$ is taken over $A \otimes A$.

Let $k' = k(\lambda)$ with $\lambda^p = a$. There exists a $k$-derivation $D$ on $k'$ such that $D(\lambda) = 1$. Let $A' = k' \otimes A$. Then $C(A') = 0$ and $P$ corresponds to an element $\sum_i \alpha_i\ D(t-d_i)/(t - d_i)$ of $L$ with $\alpha_i \in Z$ and $d_i \in k'$ such that $D(t) = t^p$ and $D(d_i) = d_i^p$ for all $i$ (cf. 6.3.1). It is easy to see that the relation

$\Delta^*(P) = q_1^*(P) \otimes q_2^*(P)$ implies

$$\sum_i \alpha_i D(t + t' - d_i)/(t + t' - d_i) - \sum_i \alpha_i D(t - d_i)/(t - d_i)$$

$$- \sum_i \alpha_i D(t' - d_i)/(t' - d_i) \in L_0(A' \otimes A'),$$

where $L_0(A' \otimes A')$ is $L_0$ for $A' \otimes A'$. Hence we have

$$\sum_i \alpha_i [\{(t + t')^{p-1} + (t + t')^{p-2} d_i + \dots + d_i^{p-1}\}$$

$$-\{t^{p-1} + t^{p-2} d_i + \dots + d_i^{p-1}\} - \{t'^{p-1} + t'^{p-2} d_i + \dots + d_i^{p-1}\}]$$

$$\in L_0(A' \otimes A').$$

Hence we have $\sum_i \alpha_i = 0$, $\sum_i \alpha_i d_i = 0, \dots, \sum_i \alpha_i d_i^{p-2} = 0$.
Since $C(A) \neq 0$, there exists an element $D(t - d)/(t - d)$ of $L$ with $d \in k'$ such that $D(d) = d^p$. Therefore $\mathrm{Ext}^1_{cent}(G, G_m) \neq H^1(G, G_m)$.

6.13.3. THEOREM. Let $G$ be the same as in 6.13.2.

Then, the abelian group $\mathrm{Ext}_{\mathrm{cent}}(G, G_m)$ is isomorphic with the following additive group:

$$\{(c_0,\ldots,c_{p-2}) \in k^{p-1} \mid c_{p-2} = c_0{}^p + c_1{}^p a + \ldots + c_{p-2}{}^p a^{p-2}\}$$

(the addition is made component-wise).

Proof. As is well-known, the Hopf algebras of $G_m$ and $G_a$ are respectively $k[u,u^{-1}]$ and $k[z]$ with the comultiplications

$$\Delta(u) = u \otimes u \quad \text{and} \quad \Delta(z) = z \otimes 1 + 1 \otimes z.$$

If we put $k' = k(\lambda)$ with $\lambda = a^{1/p}$, then $G$ is a $(k'/k)$-form of $G_a$ and the corresponding k-derivation $D_1$ of the k'-Hopf algebra $k'[z]$ (in the sense of Lemma 6.1) is defined by

$$D_1(z) = z^p \quad \text{and} \quad D_1(\lambda) = 1.$$

(The identification is $x = z^p$, $y = z - \lambda z^p$ and $O(G) = k[x,y]$). Let $D_0$ denote the trivial k-derivation of $k'[u,u^{-1}]$, i. e.,

$$D_0(u) = 0 \quad \text{and} \quad D_0(\lambda) = 1.$$

For each element $\xi \in k'$, we define a k-derivation $D_\xi$ of $k'[z,u,u^{-1}] = k'[z] \otimes_{k'} k'[u,u^{-1}]$ by

$$D_\xi(\lambda) = 1, \; D_\xi(z) = z^p \quad \text{and} \quad D_\xi(u) = \xi z u.$$

Lemma 6.13.4 below shows that $(D_\xi)^p = 0$ if and only if

$\xi^p = d^{p-2}(\xi)$, where $d$ denotes the unique k-derivation of $k'$ such that $d(\lambda) = 1$ (see 6.1). We denote by $\Xi$ the additive group of $\xi \in k'$ such that $\xi^p = d^{p-2}(\xi)$. For each $\xi \in \Xi$, put $A_\xi = k'[z,u,u^{-1}]^{D_\xi}$. This is a Hopf algebra $(k'/k)$-form of $k[z,u,u^{-1}]$, because the k-derivation $D_\xi$ commutes with the comultiplication of $k'[z,u,u^{-1}]$ in the following sense:

$$\Delta(D_\xi(c)) = \sum_{(c)} D_\xi(c_{(1)}) \otimes c_{(2)} + \sum_{(c)} c_{(1)} \otimes D_\xi(c_{(2)})$$

$$\text{for all } c \in k'[z,u,u^{-1}].$$

Observing the following commutative diagram:

$$\begin{array}{ccccc} k'[z] & \xrightarrow{i} & k'[z,u,u^{-1}] & \xrightarrow{p} & k'[u,u^{-1}] \\ \downarrow{\scriptstyle D_1} & & \downarrow{\scriptstyle D_\xi} & & \downarrow{\scriptstyle D_0} \\ k'[z] & \xrightarrow{i} & k'[z,u,u^{-1}] & \xrightarrow{p} & k'[u,u^{-1}] \end{array}$$

where $i$ and $p$ denote the canonical inclusion and projection respectively, one concludes that the homomorphisms $i$ and $p$ induce the following maps of k-Hopf algebras:

$$O(G) = k[x,y] = k'[z]^{D_1} \xrightarrow{i} A_\xi \quad \text{and}$$

$$A_\xi \xrightarrow{p} k'[u,u^{-1}]^{D_0} = k[u,u^{-1}].$$

The corresponding sequence of commutative k-group schemes

$$(E_\xi) \qquad 1 \to G_m \xrightarrow{\mathrm{Spec}(p)} \mathrm{Spec}(A_\xi) \xrightarrow{\mathrm{Spec}(i)} G \to 1$$

is clearly exact, because $(E_\xi) \otimes k'$ is split exact. Thus we obtain a map

$$\phi: \Xi \to \mathrm{Ext}_{\mathrm{cent}}(G,G_m),\ \xi \mapsto \phi(\xi) = \mathrm{class}(E_\xi).$$

It is an easy exercise to check that the map $\phi$ is a homomorphism of abelian groups. We claim that $\phi$ is an isomorphism.

The injectivity. Suppose that $\phi(\xi) = 0$ or equivalently that the sequence $(E_\xi)$ splits. This means that the canonical map

$$p: A_\xi \longrightarrow k[u,u^{-1}]$$

induces a bijection of the sets of group-like elements and hence in particular that $u \in A_\xi = k'[z,u,u^{-1}]^{D_\xi}$. Since $D_\xi(u) = \xi zu$, it follows that $\xi = 0$. This proves the injectivity of $\phi$.

The surjectivity. Let

$$(E) \qquad 1 \longrightarrow G_m \longrightarrow E \longrightarrow G \longrightarrow 1$$

be a central extension of $G$ by $G_m$. Since $\mathrm{Ext}_{\mathrm{cent}}(G_a,G_m) = 0$ by DG-III,§6,5.2, it follows that the sequence

$$(E)_{k'} \qquad 1 \to (G_m)_{k'} \to E_{k'} \to G_{k'} = (G_a)_{k'} \to 1$$

is split exact. Hence we can identify

$$k' \otimes O(E) = k'[z,u,u^{-1}].$$

Let $D$ denote the k-derivation of $k'[z,u,u^{-1}]$ associated with the k-form $O(E)$. Then we have clearly

$$D(\lambda) = 1 \quad \text{and} \quad D(z) = z^p.$$

Since $D$ should commute with the comultiplication $\Delta$ of

$k'[z,u,u^{-1}]$, it follows that

$$\Delta(D(u)) = D(u) \otimes u + u \otimes D(u)$$

or equivalently that $u^{-1}D(u)$ is a primitive element of $k'[z,u,u^{-1}]$. Since every primitive element of $k'[z,u,u^{-1}]$ is of the form $\alpha z$ with $\alpha \in k'[F]$, it follows that $u^{-1}D(u) = \alpha z$ for some $\alpha \in k'[F]$. Since $D^p = 0$, it follows easily that the degree of $\alpha$ is zero, i.e. $\alpha \in k'$. Hence we have

$$D = D_\alpha .$$

Since $(D_\alpha)^p = 0$, it follows from Lemma 6.13.4 that $\alpha \in \Xi$. This means that $\text{class}(E) = \text{class}(E_\alpha)$ and proves the surjectivity of $\phi$.

Thus we have constructed an isomorphism of abelian groups

$$\phi\colon \Xi \xrightarrow{\simeq} \text{Ext}_{\text{cent}}(G, G_m).$$

If we write an element $\xi \in k'$ as $\xi = \sum_{i=0}^{p-1} c_i\lambda^i$ with $c_i \in k$, then $\xi^p = \sum_{i=0}^{p-1} c_i{}^p a^i$ and $d^{p-2}(\xi) = (p-2)!c_{p-2} + (p-1)!c_{p-1}\lambda = c_{p-2} - c_{p-1}\lambda$. Hence $\xi \in \Xi$ if and only if

$$c_{p-1} = 0 \quad \text{and} \quad c_{p-2} = c_0{}^p + c_1{}^p a + \dots + c_{p-2}{}^p a^{p-2}.$$

This proves Theorem.

6.13. 4. LEMMA. <u>For each</u> $\xi \in k'$, <u>let</u> $D_\xi$ <u>denote the</u> k-<u>derivation of</u> $k'[z,u,u^{-1}]$ <u>defined by</u>

$$D_\xi(\lambda) = 1,\ D_\xi(u) = \xi z u \quad \text{and} \quad D_\xi(z) = z^p.$$

<u>Then we have</u>

$$(D_\xi)^p(u) = d^{p-1}(\xi)zu + (\xi^p - d^{p-2}(\xi))z^p u.$$

<u>where</u> $d$ <u>is the canonical</u> k-<u>derivation of</u> $k'$. <u>Hence</u> $(D_\xi)^p = 0$ <u>if and only if</u> $\xi^p = d^{p-2}(\xi)$.

<u>Proof</u>. Put $D = D_\xi$. Since $D^p$ clearly commutes with the comultiplication $\Delta$ of $k'[z,u,u^{-1}]$, it follows that

$$u^{-1}D^p(u) = \alpha z \quad \text{for some} \quad \alpha \in k'[F].$$

(See the proof above). On the other hand $u^{-1}D^n(u)$ belongs to $k'[z]$ for all $n > 0$ and one can prove by induction that $\deg_z(u^{-1}D^n(u)) \leqq (p-1)(n-1) + 1$ for all $n > 0$. In particular we have $\deg_z(u^{-1}D^p(u)) \leq (p-1)^2 + 1 < p^2$. Hence $\alpha$ must be of the form

$$\alpha = \eta + \eta' F \quad \text{with} \quad \eta, \eta' \in k'.$$

Next, for each integer $0 < i < p$, one can prove by induction that

$$u^{-1}D^i(u) = d^{i-1}(\xi)z + (\text{the terms of } 1 < \text{degree} < i)$$
$$+ \xi^i z^i + (i-1)d^{i-2}(\xi)z^p + (\text{the terms of degree} > p),$$

where the degree is taken with respect to $z$. In particular putting $i = p-1$, we can determine the coefficients of $zu$ and $z^p u$ in $D^p(u) = D(D^{p-1}(u))$ to be $\eta = d^{p-1}(\xi)$ and $\eta' = \xi^p - d^{p-2}(\xi)$ respectively. This proves Lemma.

## 7. Actions of unipotent group schemes

*In this section (§7), the ground field k is arbitrary.*

7.0. *We consider the following standard situation, for which the notations will be fixed all through* §7:

7.0.1. An affine algebraic k-group scheme G acts on an affine k-scheme X = Spec(A) on the right through

$$u: X \times G \longrightarrow X .$$

The coaction determining u is written as

$$\rho: A \longrightarrow A \otimes O(G)$$

which is a k-algebra homomorphism giving a right O(G)-comodule structure to A.

7.0.2. The underlying k-module of A can be viewed as a left k-G-module; denote by $H^n(G,A)$ the n-th Hochschild cohomology group of G with coefficients in A — see 7.2 below and DG - II, §3.

7.0.3. Let $\tilde{\tilde{\mathbb{C}}}_k$ denote the category of all set-valued (fpqc)-sheaves on $\mathbb{Alg}_k$. (This is what is referred to as *faisceau dur* in DG - III, §2, 1.3. Thus, in the fashion of DG, one would write $(\mathbb{Alg}_k \mathbb{E})^{\approx}$ in lieu of $\tilde{\tilde{\mathbb{C}}}_k$.) We consider the quotient sheaf of X by the action of G, denoted by $X\tilde{\tilde{/}}G$.

7.0.4. In a like manner, we denote by $\boxed{C}_k$, $\boxed{C}_k^{\sim}$ respectively the category of all set-valued presheaves (covariant functors) and that of all set-valued (fppf)-sheaves on $\boxed{Alg}_k$. Thus, $\boxed{C}_k = \boxed{Alg}_k\boxed{E}$ and $\boxed{C}_k^{\sim} = (\boxed{Alg}_k\boxed{E})^{\sim}$. Note that the inclusions $\boxed{Aff}.\boxed{Sch}_k \subset \boxed{Sch}_k \subset \boxed{C}_k^{\approx} \subset \boxed{C}_k^{\sim} \subset \boxed{C}_k$ hold.

7.1. The purpose of the present section is to prove the following

7.1.1. THEOREM. *With the notations and assumptions of* 7.0, *assume further that* $G$ *is unipotent. Then, the following statements are equivalent to each other:*

(i) *The action of* $G$ *on* $X$ *is free* (see 7.1 for the definition), *and* $X\tilde{\tilde{/}}G$ *is representable and affine;*

(ii) $H^n(G,A) = 0$ *for all* $n \geq 1$; *and*

(iii) $H^1(G,A) = 0$ .

7.1.2. *Comments.* Earlier, one of the authors asserted a result [4; Theorem 1] roughly the same as 7.1.1 here, somewhat stronger in some ways and weaker in others. His proofs of the assertion and a key lemma for it (*ibid.*, Th.1 and Lemma 3) tended to go over the ground too fast. A careful re-examination of the arguments in [4] has revealed that unfortunately certain of them contain gaps that are hard to fill; as a result, both Theorem 1 and Lemma 3 of [4] remain unsupported by proof *in the stated form*. Rather than patching up and making corrections here and there on the basis of [4], we have preferred to rebuild the theorem from the ground up, in the framework best suited for our purpose in the present work (cf. §8 below). We point out, however, that the basic ideas of proof of 7.1.1 remain the same as those of [4]. The said ideas may be summarized as follows: The questions as to whether or not the action is free and whether or not the quotient is affine representable are indifferent to scalar field extensions. The Demazure-Hochschild cohomology commutes with scalar field extensions. Therefore, in proving 7.1.1 one may assume $G$ to be k-solvable with a central series in which each successive quotient is either $G_a$ or $\alpha_p$. By interpreting the $G_a$- or $\alpha_p$-action through higher derivations, one proves 7.1.1 for this case directly. The spectral sequence (ibid., Lemma 3) takes care of $G$ which is a multiple extension of $G_a$'s and $\alpha_p$'s.

7.2. Let $G$ be a k-group functor. By a G-module functor we mean a commutative k-group functor $M$ with an operation of $G$, denoted $(g,m) \mapsto {}^g m$, such that ${}^g(m+m') = {}^g m + {}^g m'$ for all $R \in \boxed{Alg}_k$, $g \in G(R)$, $m, m' \in M(R)$ (DG-II,§3). By a k-G-module we mean a pair $(V,\rho)$ where $V$ is a k-module and $\rho: G \to \underline{\underline{GL}}(V)$ a linear representation. Thus a k-G-module is nothing else than a G-module functor of the form $V_{\underline{\underline{a}}}$ such that for any $R \in \boxed{Alg}_k$ and $g \in G(R)$ the induced endomorphism of $V \otimes R$ by $g$ is R-linear (DG-II,§2).

Suppose that $G$ is an affine k-group scheme. As usual,

the Hopf algebra of $G$ will be denoted by $O(G)$. As is well-known (DG - II, §2, 2.1), the linear representations $G \to \underline{\underline{GL}}(V)$ correspond bijectively with the right comodule structures $V \to V \otimes O(G)$. Hence we may and shall identify the category of k-G-modules with that of right $O(G)$-comodules, denoted $\underline{\text{Comod}}_{O(G)}$, if $G$ is affine.

Let $M$ be a G-module functor. The Hochschild cohomology groups $H_0^n(G,M)$ of $G$ with coefficients in $M$ are defined as follows (DG - II, §3): Put

$$C^n(G,M) := \underline{C}_k(G^n,M) \quad \text{for } n \geq 0.$$

Define the coboundary homomorphism $\partial^n\colon C^n(G,M) \to C^{n+1}(G,M)$ by $\partial^n := \sum_{i=0}^{n+1}(-1)^i \partial_i^n$, where for all $g_1,\dots,g_{n+1} \in G(R)$ and $f \in C^n(G,M)$

$$\begin{cases} (\partial_0^n f_R)(g_1,\dots,g_{n+1}) := {}^{g_1}f_R(g_2,\dots,g_{n+1}) \,, \\ (\partial_i^n f_R)(g_1,\dots,g_{n+1}) := f_R(g_1,\dots,g_i g_{i+1},\dots,g_{n+1}),\ 1 \leq i \leq n \,, \\ (\partial_{n+1}^n f_R)(g_1,\dots,g_{n+1}) := f_R(g_1,\dots,g_n) \,. \end{cases}$$

Then $\partial^n \partial^{n-1} = 0$ for $n > 0$ and the n-th cohomology group of the complex $C^{\cdot}(G,M) := ((C^n(G,M)),(\partial^n))$ is denoted by $H_0^n(G,M)$ and called <u>the</u> n-<u>th</u> <u>Hochschild</u> <u>cohomology</u> <u>group</u> <u>of</u> $G$ <u>with</u>

<u>coefficients</u> <u>in</u> M.

Let V be a k-G-module. Then the group $H_0^n(G,V_{\underline{\underline{a}}})$ will be denoted by $H^n(G,V)$. Suppose that G is affine: $G = \mathrm{Spec}\, O(G)$. Let $\rho\colon V \to V \otimes O(G)$ be the associated comodule structure map. Let $C^\cdot(G,V)$ be the following complex:

$$C^n(G,V) := V \otimes O(G) \otimes \cdots \otimes O(G) \quad (\text{n copies of } O(G)),$$

$$\partial^n := \textstyle\sum_{i=0}^{n+1} (-1)^i \partial_i^n \quad ,$$

where $\partial_i^n\colon V \otimes O(G)^{\otimes n} \to V \otimes O(G)^{\otimes(n+1)}$ is defined by

$$\begin{cases} \partial_0^n(v \otimes a_1 \otimes \ldots \otimes a_n) := \rho(v) \otimes a_1 \otimes \ldots \otimes a_n \ , \\ \partial_i^n(v \otimes a_1 \otimes \ldots \otimes a_n) := v \otimes a_1 \otimes \ldots \otimes \Delta a_i \otimes \ldots \otimes a_n, \ 1 \leq i \leq n \ , \\ \partial_{n+1}^n(v \otimes a_1 \otimes \ldots \otimes a_n) := v \otimes a_1 \otimes \ldots \otimes a_n \otimes 1 \ , \end{cases}$$

where $\Delta\colon O(G) \to O(G) \otimes O(G)$ denotes the comultiplication. Then we have $C^\cdot(G,V_{\underline{\underline{a}}}) \simeq C^\cdot(G,V)$ (see DG-II, §3, 3.1). Hence $H^n(G,V) \simeq H^n(C^\cdot(G,V))$ in case G is affine.

7.3. Let G be a k-group functor and K a normal subgroup functor of G. By K normal we mean that K(R) is a normal subgroup of G(R) for all $R \in \boxed{\mathrm{Alg}}_k$. Let M be a G-module functor. For an n-cochain $f \in C^n(K,M)$ and an element $x \in G(k)$, let the n-cochain $x \rightharpoonup f \in C^n(K,M)$ be defined by

$$[x \rightharpoonup f](g_1,\ldots,g_n) = {}^x f(x^{-1}g_1x,\ldots,x^{-1}g_nx),$$

$$\text{for all} \quad g_1,\ldots,g_n \in K(R), \quad R \in \underline{\mathrm{Alg}}_k .$$

Then one can easily verify that

$$\partial^n(x \rightharpoonup f) = x \rightharpoonup \partial^n f .$$

Hence the left action of $G(k)$ on $C^{\cdot}(K,M)$ induces a left action of $G(k)$ on $H_0^n(K,M)$.

If $f\colon M \to N$ is a homomorphism of G-module functors, then the induced homomorphisms of groups $H_0^n(K,f)\colon H_0^n(K,M) \to H_0^n(K,N)$ are easily seen to commute with the $G(k)$-action, since the homomorphism of complexes $C^{\cdot}(K,f)\colon C^{\cdot}(K,M) \to C^{\cdot}(K,N)$ already commutes with the $G(k)$-action.

Suppose that $K$ is affine. Then a short exact sequence

$$0 \longrightarrow M' \longrightarrow M \longrightarrow M'' \longrightarrow 0$$

of G-module functors induces an exact sequence of complexes

$$0 \longrightarrow C^{\cdot}(K,M') \longrightarrow C^{\cdot}(K,M) \longrightarrow C^{\cdot}(K,M'') \longrightarrow 0 .$$

Since each element $x$ of $G(k)$ induces a homomorphism of exact sequences

$$\begin{array}{ccccccccc} 0 & \longrightarrow & C^{\cdot}(K,M') & \longrightarrow & C^{\cdot}(K,M) & \longrightarrow & C^{\cdot}(K,M'') & \longrightarrow & 0 \\ & & {\scriptstyle x \rightharpoonup -}\downarrow & & {\scriptstyle x \rightharpoonup -}\downarrow & & {\scriptstyle x \rightharpoonup -}\downarrow & & \\ 0 & \longrightarrow & C^{\cdot}(K,M') & \longrightarrow & C^{\cdot}(K,M) & \longrightarrow & C^{\cdot}(K,M'') & \longrightarrow & 0 \end{array},$$

it follows that the connecting homomorphism $H_0^n(K,M'') \longrightarrow H_0^{n+1}(K,M')$ also commutes with the $G(k)$-action.

7.3.1. LEMMA. _If_ $K$ _is affine, the induced action of_ $K(k)$ _on_ $H_0^n(K,M)$ _is trivial for every_ $G$-_module functor_ $M$.

_Proof_. We can assume that $G = K$. The lemma is obvious for $n = 0$, since $H_0^0(K,M) = M^K(k)$ . Notice that the functor $H_0^{\cdot}(K,-)$ is effaceable (cf. the proof of DG-II, §3, 1.3). Hence we can form a short exact sequence of $K$-module functors

$$0 \longrightarrow M \longrightarrow \overline{M} \longrightarrow M'' \longrightarrow 0$$

such that $H_0^n(K,\overline{M}) = 0$ for all $n > 0$. Then we have $H^{n-1}(K,M'') \longrightarrow H^n(K,M) \longrightarrow 0$ exact for all $n > 0$ , which commutes with the $K(k)$-action. Therefore the assertion holds by the induction argument.

7.4. We place ourselves in the situation of 7.0 above, possessing a right $G$-action on $X = \mathrm{Spec}(A)$. In this situation $A$ will be referred to as a right $O(G)$-comodule algebra. The action of $G$ on $X$ is said to be _free_ if the morphism $(pr_1, u)\colon X \times G \longrightarrow X \times X$, given by $(x,g) \longmapsto (x,xg)$ for all $x \in X(R)$, $g \in G(R)$, is a monomorphism of k-functors, where $xg = u(x,g)$ and $pr_1\colon X \times G \to X$ is the first projection. The quotient sheaf $X\tilde{\tilde{/}}G$ (see 7.0.3) is defined by the following exact sequence in $\tilde{\tilde{\boxed{C}}}_k$ (cf. DG-III, §2, 1.3):

$$X \times G \underset{u}{\overset{pr_1}{\rightrightarrows}} X \longrightarrow X\tilde{\tilde{/}}G \ .$$

If the action is free we have clearly

$$X \times G \xrightarrow{\simeq} X \times_{X/G} X \ .$$

Suppose that $X\tilde{\tilde{/}}G$ is representable affine and put $X\tilde{\tilde{/}}G = \mathrm{Spec}(B)$. Then $A$ is a faithfully flat $B$-algebra by DG-III, §1, 2.11 and also III, §3, 2.5. It is easy to see that $B = A^G$ (taken with respect to the induced k-G-module structure on A). Therefore the following are equivalent:

(i) The action $u$ is free and $X\tilde{\tilde{/}}G$ is affine.

(ii) $A$ is a faithfully flat $A^G$-algebra and the map

$$A \otimes_{A^G} A \longrightarrow A \otimes O(G),\ a \otimes b \longmapsto (a \otimes 1)\rho(b)$$

is an isomorphism.

7.4.1. LEMMA. *If one of the equivalent conditions above holds, then* $H^n(G,A) = 0$ *for all* $n > 0$.

*Proof.* Indeed we have an isomorphism of k-functors

$$X \times G^n \xrightarrow{\simeq} X \times_{X\tilde{\tilde{/}}G} \cdots \times_{X\tilde{\tilde{/}}G} X \quad (n+1 \text{ copies of } X)$$

$$(x,g_1,\ldots,g_n) \longmapsto (x,xg_1,xg_1g_2,\ldots,xg_1\cdots g_n).$$

Let

$$A \otimes O(G)^{\otimes n} \xleftarrow{\simeq} A \otimes_{A^G} \cdots \otimes_{A^G} A \quad (n+1 \text{ copies of } A)$$

be the associated isomorphism of k-algebras. These homomorphisms form an isomorphism of complexes

$$C^{\cdot}(G,A) \simeq C^{\cdot}(A/A^G, G_a)$$

the right-hand side of which denoting the Amitsur complex relative to $G_a$ and the canonical projection $\mathrm{Spec}(A) \to \mathrm{Spec}(A^G)$ (cf. DG-III, §5, 5.2 and §4,6.4). Since $H^n(A/A^G,G_a) = 0$ for $n > 0$ by DG-I, §1,2.7, the assertion follows.

7.5. Let $G$ be an affine k-group scheme and $K$ a closed normal subgroup scheme of $G$. Let $\pi\colon O(G) \to O(K)$ be the canonical Hopf algebra surjection. Let $V$ be a k-vector space. The composite

$$V \otimes O(G) \xrightarrow{1 \otimes \Delta} V \otimes O(G) \otimes O(G) \xrightarrow{1 \otimes 1 \otimes \pi} V \otimes O(G) \otimes O(K)$$

makes $V \otimes O(G)$ into a right $O(K)$-comodule (i.e., a k-K-module).

7.5.1. COROLLARY. *With the above k-K-module structure on* $V \otimes O(G)$, *we have*

$$H^n(K, V \otimes O(G)) = 0 \quad \textit{for all} \quad n > 0 .$$

*Proof*. Since $C^{\cdot}(K, V \otimes O(G)) \simeq V \otimes C^{\cdot}(K, O(G))$, it suffices to show $H^n(K, O(G)) = 0$ for $n > 0$. Since the k-K-module structure on $O(G)$ is induced from the right action (which is free)

$$G \times K \longrightarrow G, \ (g,h) \longmapsto gh$$

the assertion follows from 7.4.1, in view of the fact that $G \tilde{\tilde{/}} K$ is affine (DG-III, §3,7.2).

7.6. Let $G$ be an affine k-group scheme and $K$ a closed normal subgroup scheme of $G$. We shall now determine the right derived functors of

$$\boxed{\text{Comod}}_{O(G)} \longrightarrow \boxed{\text{Comod}}_{O(G\tilde{\tilde{/}}K)} , \text{ given by } V \longmapsto V^K ,$$

where one should notice that $V^K$ is naturally a k-$G\tilde{\tilde{/}}K$-module for any k-G-module $V$.

7.6.1. Our task begins with making $H^n(K,V)$ into k-$\tilde{\tilde{G/K}}$-modules for each k-G-module V. Let V be a k-G-module. For each k-algebra R, $V\otimes R$ is an R-$(G\otimes R)$-module. Hence a natural G(R)-action (from the left) on $H^n(K\otimes R, V\otimes R)$ is defined as in 7.3. Since for each $\lambda \in R$ the induced map $1\otimes\lambda: V\otimes R \longrightarrow V\otimes R$ is an R-$(G\otimes R)$-module homomorphism, the G(R)-action on $H^n(K\otimes R, V\otimes R)$ is R-linear. Note that

$$H^n(K\otimes R, V\otimes R) \simeq H^n(K,V)\otimes R$$

(DG-II, §3, 3.6). Hence G(R) acts on $H^n(K,V)\otimes R$ from the left R-linearly. Since this action is natural with respect to R, we have defined a k-G-module structure on $H^n(K,V)$. Since the induced K(R)-action on $H^n(K,V)\otimes R$ is trivial by 7.3.1, $H^n(K,V)$ becomes a k-$\tilde{\tilde{G/K}}$-module.

7.6.2. Let $f: V \to V'$ be a k-G-module map. Since $f_R: V\otimes R \to V'\otimes R$ is an R-$(G\otimes R)$-module map for all R, the induced homomorphisms $H^n(K\otimes R, f_R): H^n(K,V)\otimes R \to H^n(K,V')\otimes R$ commute with the G(R)-action. Hence the maps $H^n(K,f)$: $H^n(K,V) \to H^n(K,V')$ are k-$\tilde{\tilde{G/K}}$-module maps. Let $0 \to V' \to V \to V'' \to 0$ be a short exact sequence of k-G-modules. Since $0 \to V'\otimes R \to V\otimes R \longrightarrow V''\otimes R \to 0$ is exact for every $R \in \mathrm{Alg}_k$, it follows that the connecting homomorphism $H^n(K,V'')\otimes R \longrightarrow H^{n+1}(K,V')\otimes R$ commutes with the G(R)-action. Therefore the resulting long exact sequence

$$\cdots \longrightarrow H^n(K,V) \longrightarrow H^n(K,V'') \longrightarrow H^{n+1}(K,V') \longrightarrow H^{n+1}(K,V) \longrightarrow \cdots$$

consists of k-$G\widetilde{/}K$-module maps. Thus the functors $H^n(K,-)$: $\boxed{\text{Comod}}_{O(G)} \longrightarrow \boxed{\text{Comod}}_{O(G\widetilde{\widetilde{/}}K)}$ constitute a cohomological functor (i.e., an exact connected sequence of functors), which we shall denote by $H^{\cdot}(K,-)$.

7.6.3. PROPOSITION. The (cohomological) functor $H^{\cdot}(K,-)$: $\boxed{\text{Comod}}_{O(G)} \longrightarrow \boxed{\text{Comod}}_{O(G\widetilde{\widetilde{/}}K)}$ is the right derived (cohomological) functor of the functor $\boxed{\text{Comod}}_{O(G)} \longrightarrow \boxed{\text{Comod}}_{O(G\widetilde{\widetilde{/}}K)}$ given by $V \mapsto V^K$.

Proof. Let $V$ be a k-G-module. Then the $O(G)$-comodule structure map of $V$: $V \to V \otimes O(G)$ is injective and k-G-linear, where the k-G-module structure on $V \otimes O(G)$ is the canonical one. Since $H^n(K, V \otimes O(G)) = 0$ for $n > 0$ by 7.5.1, it follows that the functor $H^{\cdot}(K,-)$ is effaceable and hence the right derived functor of $H^0(K,-) = -^K$.

7.7. Let $G$ be an affine k-group scheme and $K$ a closed normal subgroup scheme of $G$. The functor $V \mapsto V^G$: $\boxed{\text{Comod}}_{O(G)} \longrightarrow \boxed{\text{Mod}}_k$ factors as

$$\boxed{\text{Comod}}_{O(G)} \xrightarrow{\ \Psi\ } \boxed{\text{Comod}}_{O(G\widetilde{\widetilde{/}}K)} \xrightarrow{\ \Phi\ } \boxed{\text{Mod}}_k$$

where $\Psi(V) = V^K$ and $\Phi(W) = W^{G\widetilde{\widetilde{/}}K}$. $\Phi$ and $\Psi$ are clearly left exact. The right derived functors of $\Phi$, $\Psi$ and $\Phi\Psi$ are respectively $H^\cdot(G/K,-)$, $H^\cdot(K,-)$ and $H^\cdot(G,-)$ by 7.6.3.

7.7.1. PROPOSITION (cf. [4;Lemma3], 7.1.2 above and DG-III,§6,3.3). <u>There is a spectral sequence</u>

$$H^p(G\tilde{\tilde{/}}K, H^q(K,V)) \Longrightarrow H^{p+q}(G,V)$$

<u>for every</u> k-G-<u>module</u> V.

<u>Proof</u>. It suffices to show that the abelian categories $\boxed{\text{Comod}}_{O(G)}$ and $\boxed{\text{Comod}}_{O(G\tilde{\tilde{/}}K)}$ have enough injectives and the functor $\Psi$ preserves them (cf. [17;Th.2.4.1]). For any k-vector space W, $W \otimes O(G)$ has a natural right $O(G)$-comodule structure $1 \otimes \Delta: W \otimes O(G) \longrightarrow W \otimes O(G) \otimes O(G)$. With this structure $W \otimes O(G)$ is an injective k-G-module, since

$$\boxed{\text{Comod}}_{O(G)}(V, W \otimes O(G)) \simeq \boxed{\text{Mod}}_k(V,W)$$

for every k-G-module V. Since for any k-G-module V, the $O(G)$-comodule structure map $\rho: V \to V \otimes O(G)$ is k-G-linear, the abelian category $\boxed{\text{Comod}}_{O(G)}$ has enough injectives. Since $\Psi(W \otimes O(G)) = (W \otimes O(G))^K = W \otimes O(G)^K = W \otimes O(G\tilde{\tilde{/}}K)$ (see 7.4) for any vector space W, the functor $\Psi$ preserves injectives. This proves our proposition.

7.7.2. COROLLARY. Let $G$ be an affine k-group scheme and $K$ a closed normal subgroup scheme of $G$. For every k-G-module $V$, we have an exact sequence of k-vector spaces

$$0 \longrightarrow H^1(G\tilde{/}K, V^K) \longrightarrow H^1(G,V) \longrightarrow H^1(K,V)^G \longrightarrow H^2(G\tilde{/}K, V^K) \longrightarrow H^2(G,V) .$$

For a proof consult Cartan-Eilenberg's textbook. (esp. Th.5.12, p.328).

7.8. Let $G$ be an affine algebraic unipotent k-group shceme. (Thus, the Hopf algebra $O(G)$ is irreducible and finitely generated.) Let $X = \mathrm{Spec}(A)$ be an affine k-scheme, and suppose that a right action $u: X \times G \to X$ of $G$ is given. As in 7.0, let $\rho: A \to A \otimes O(G)$ be the associated right comodule algebra structure map. As announced in 7.1, the purpose of this section is to prove 7.1.1:

THEOREM (cf. [4; Th.1] and 7.1.2 above). In the situation of 7.8, the following are equivalent:

(i) The action of $G$ is free and the (fpqc)-quotient k-sheaf $X\tilde{\tilde{/}}G$ is affine representable.

(ii) $H^n(G,A) = 0$ for all $n > 0$.

(iii) $H^1(G,A) = 0$.

Proof. (i) $\Longrightarrow$ (ii) follows from 7.4.1, and (ii) $\Longrightarrow$ (iii) is clear. It remains to prove (iii) $\Longrightarrow$ (i). Since $G$ is algebraic unipotent, there is a central series $G = G_0 \supset G_1 \ldots \supset G_n = (1)$ of closed subgroup schemes of $G$ such that $G_{i-1}/G_i$ are isomorphic to some closed subgroup schemes of $G_a$ (see, e.g., DG-IV,§2,2.5). By the induction argument, we can assume that the theorem holds true for the group $G/K$, where $K = G_{n-1}$. Suppose that $H^1(G,A) = 0$. Consider the following exact sequence:

$$0 \longrightarrow H^1(G\tilde{/}K,A^K) \longrightarrow H^1(G,A) \longrightarrow H^1(K,A)^G \longrightarrow H^2(G\tilde{/}K,A^K) ,$$
$$\begin{array}{c} \| \\ 0 \end{array}$$

where $G\tilde{/}K$ denotes $G/K$ considered canonically as an (fppf)-sheaf. Since $H^1(G\tilde{/}K,A^K) = 0$, it follows that $H^2(G\tilde{/}K,A^K) = 0$. (Apply (iii) $\Longrightarrow$ (ii) to $G\tilde{/}K$.) Hence $H^1(K,A)^G = 0$. Let $x$ be a primitive element of $O(K)$ (viz., $\Delta(x) = x \otimes 1 + 1 \otimes x$). The element $1 \otimes x$ of $A \otimes O(K)$ is easily seen to be a 1-cocycle of the complex $C^{\cdot}(K,A)$ (see 7.9 below). Let $f_x$: $K \rightarrow A_a$ be the 1-cocycle associated with $1 \otimes x$ through the canonical isomorphism $C^{\cdot}(K,A) \simeq C^{\cdot}(K,A_{\underline{\underline{a}}})$ in 7.2. Since $K$ is central in $G$, $f_x$ factors through $k_{\underline{\underline{a}}} \rightarrow A_{\underline{\underline{a}}}$, $1 \mapsto 1$, and $G$ acts on $k_a$ trivially, it follows that the 1-cocycle $f_x$ is G-invariant. Since $H^1(K,A)^G = 0$ , there is an element $a_x \in A$ such that

$$\rho(a_x) - a_x \otimes 1 = 1 \otimes x \quad \text{in} \quad A \otimes O(K).$$

Since $K$ is isomorphic to some subgroup scheme of $G_a$, the Hopf algebra $O(K)$ is generated by the primitive elements. Proposition 7.9.1 below implies that the action $X \times K \to X$ is free and that $X\tilde{\tilde{/}}K$ is affine. Since $X\tilde{\tilde{/}}K = \mathrm{Spec}(A^K)$ and $H^1(G\tilde{/}K, A^K) = 0$, it follows from the induction hypothesis that the induced action

$$X\tilde{\tilde{/}}K \times G\tilde{/}K \longrightarrow X\tilde{\tilde{/}}K$$

is free and that the quotient $(X\tilde{\tilde{/}}K)\tilde{\tilde{/}}(G\tilde{/}K) \simeq X\tilde{\tilde{/}}G$ is affine. Consider the map $X \times G \longrightarrow X \times X$, $(x,g) \longmapsto (x,xg)$. If $xg = xh$ for some $x \in X(R)$, $g, h \in G(R)$, we have $gh^{-1} \in K(R)$, since $G\tilde{/}K$ acts on $X\tilde{\tilde{/}}K$ freely. Since $\pi(gh^{-1}) = \lambda$ and $K$ acts on $X$ feely, it follows that $gh^{-1} = 1$. This proves that the action of $G$ on $X$ is free and completes the proof of (iii) $\Longrightarrow$ (i).

7.9. It remains to prove our Theorem in the case where $O(G)$ is generated by the primitive elements. We shall treat this case using the theory of Hopf algebras. Let $H$ be a commutative Hopf algebra. Let $P(H)$ be the set of primitive elements in $H$. Thus $P(H) = \{x \in H \mid \Delta(x) = x \otimes 1 + 1 \otimes x\}$. Let $A$ be a right commutative $H$-comodule algebra. This means that an algebra map

$$\rho\colon A \longrightarrow A \otimes H$$

which is also a right $H$-comodule structure map is given. Put

$$A^H = \{a \in A \mid \rho(a) = a \otimes 1\} .$$

It follows from 7.4 that the following are equivalent:

(i) The right action $\mathrm{Spec}(\rho)\colon \mathrm{Spec}(A) \times \mathrm{Spec}(H) \to \mathrm{Spec}(A)$ is free and $\mathrm{Spec}(A)\tilde{/}\mathrm{Spec}(H)$ is affine.

(ii) $A$ is a faithfully flat $A^H$-algebra and we have an isomorphism of k-algebras

$$A \otimes_{A^H} A \xrightarrow{\simeq} A \otimes H , \quad a \otimes b \longmapsto (a \otimes 1)\rho(b) .$$

Recall that the Hochschild complex $C^{\cdot}(\mathrm{Spec}(H),A)$ goes as follows in lower dimensions.

$$\begin{array}{ccccc} C^0(\mathrm{Spec}(H),A) & \xrightarrow{\partial^0} & C^1(\mathrm{Spec}(H),A) & \xrightarrow{\partial^1} & C^2(\mathrm{Spec}(H),A) \\ \| & & \| & & \| \\ A & & A \otimes H & & A \otimes H \otimes H \end{array} ,$$

$$\partial^0(a) = \rho(a) - a \otimes 1 ,$$

$$\partial^1(a \otimes h) = \rho(a) \otimes h - a \otimes \Delta(h) + a \otimes h \otimes 1.$$

It follows that $\partial^1(1 \otimes x) = 0$ for all $x \in P(H)$.

7.9.1. PROPOSITION. _If_ $H$ _is generated by_ $P(H)$ _as an algebra, the following are equivalent:_

(i) _The right action_ $\mathrm{Spec}(\rho)\colon \mathrm{Spec}(A) \times \mathrm{Spec}(H) \to \mathrm{Spec}(A)$

<u>is free and the quotient</u> $\mathrm{Spec}(A)\tilde{/}\mathrm{Spec}(H)$ <u>is affine</u>.

(ii) <u>For each element</u> $x \in P(H)$, <u>there is an element</u> $a_x \in A$ <u>such that</u> $\partial^0(a_x) = \rho(a_x) - a_x \otimes 1 = 1 \otimes x$.

(iii) $H^n(\mathrm{Spec}(H), A) = 0$ <u>for all</u> $n > 0$.

(Notice that we do not assume $H$ to be finitely generated.)

<u>Proof</u>. (i) $\Longrightarrow$ (iii) $\Longrightarrow$ (ii) are clear. We show that (ii) $\Longrightarrow$ (i). For each element $x$ of $P(H)$ take an element $a_x$ of $A$ such that

$$\partial^0(a_x) = \rho(a_x) - a_x \otimes 1 = 1 \otimes x .$$

Notice that $a_x$ is uniquely determined <u>modulo</u> $A^H = \mathrm{Ker}(\partial^0)$. Hence the element $\phi(x) = 1 \otimes a_x - a_x \otimes 1 \in A \otimes_{A^H} A$ is well-defined. Let $A'$ be the sub-algebra of $A$ generated by $A^H$ and all $a_x$, $x \in P(H)$. Thus

$$A' = A^H[a_x;\ x \in P(H)].$$

The map $\phi: P(H) \to A' \otimes_{A^H} A'$, $x \mapsto 1 \otimes a_x - a_x \otimes 1$ is clearly k-linear. If the characteristic $p$ of $k$ is positive, we have $\phi(x^p) = \phi(x)^p$, since $\partial^0(a_x{}^p) = \partial^0(a_x)^p = 1 \otimes x^p$. It is well-known that $H$ is canonically isomorphic to $U(P(H))$, the universal enveloping algebra of the abelian p-Lie algebra $P(H)$ if $p > 0$ (cf. [13;page 274, Th.13.0.1 and page 284, Prop.13.2.3]), and the usual universal enveloping algebra of

$P(H)$ if $p = 0$. Hence the map $\phi$ can be uniquely extended to an algebra map $\phi\colon H \to A' \otimes_{A^H} A'$. Let

$$\Phi\colon A' \otimes H \longrightarrow A' \otimes_{A^H} A', \quad a \otimes h \longmapsto (a \otimes 1)\phi(h)$$

be the induced $A'$-algebra map. On the other hand, since $A'$ is clearly a sub-H-comodule, we can well define an algebra map

$$\Psi\colon A' \otimes_{A^H} A' \to A' \otimes H, \quad a \otimes b \longmapsto (a \otimes 1)\rho(b) .$$

We claim that $\Phi\Psi = 1 = \Psi\Phi$.

$\Psi\Phi = 1$: Indeed if $x \in P(H)$, then $\Psi(\Phi(1 \otimes x)) = \Psi(1 \otimes a_x - a_x \otimes 1) = \rho(a_x) - a_x \otimes 1 = 1 \otimes x$.

$\Phi\Psi = 1$: Indeed if $x \in P(H)$, then $\Phi(\Psi(1 \otimes a_x)) = \Phi(\rho(a_x)) = \Phi(a_x \otimes 1 + 1 \otimes x) = a_x \otimes 1 + 1 \otimes a_x - a_x \otimes 1 = 1 \otimes a_x$.

Let $\{e_\lambda\}_{\lambda \in \Lambda}$ be a k-basis of $P(H)$. Let $[\Lambda]$ be the set of functions $m\colon \Lambda \to \mathbb{N}$ (= the integers $\geqq 0$) such that

1) the set of $\lambda \in \Lambda$ such that $m(\lambda) \neq 0$ is finite, and
2) $m(\lambda) < p$ for all $\lambda \in \Lambda$ if $\mathrm{char}(k) = p > 0$.

For an element $m$ of $[\Lambda]$, put

$$e^m = \Pi_{\lambda \in \Lambda}\, e_\lambda^{\,m(\lambda)} .$$

(This is well-defined since H is commutative.) By the celebrated Poincaré-Birkhoff-Witt theorem, $\{e^m \mid m \in [\Lambda]\}$ forms a k-basis of H. We call $|m| = \sum m(\lambda)$ the degree of

$e^m$.

For each $\lambda \in \Lambda$, let $a_\lambda \in A$ be an element such that

$$\partial^0(a_\lambda) = \rho(a_\lambda) - a_\lambda \otimes 1 = 1 \otimes e_\lambda .$$

It is easy to see that $A'$ is generated by $\{a_\lambda\}_{\lambda\in\Lambda}$ over $A^H$ as an algebra. For an element $m \in [\Lambda]$, put $a^m = \Pi a_\lambda{}^{m(\lambda)}$.

7.9.2. LEMMA. <u>The set</u> $\{a^m\}_{m\in[\Lambda]}$ <u>forms an</u> $A^H$-<u>basis of</u> $A'$.

<u>Proof</u>. That $A'$ is generated by $\{a^m\}$ is clear if $\mathrm{char}(k) = 0$. Suppose that $\mathrm{char}(k) = p > 0$. Let $\lambda \in \Lambda$. Then $e_\lambda{}^p$ is of the form $\sum_\mu c_{\lambda\mu} e_\mu$ with $c_{\lambda\mu} \in k$. Since then $a_\lambda{}^p - \sum_\mu c_{\lambda\mu} a_\mu \in A^H$, it follows that the $A^H$-submodule $A^H + \sum_{\lambda\in\Lambda} A^H a_\lambda$ is closed under the p-power operation. This implies immediately that the $A^H$-module $A'$ is generated by $\{a_m\}$.

It remains to see the independence of $\{a_m\}$. Suppose given an $A^H$-linear relation $\sum_{m\in[\Lambda]} \xi_m a^m = 0$, where $\xi_m \in A^H$ and the set of $m \in [\Lambda]$ such that $\xi_m \neq 0$ is finite. Let $n$ be the highest degree of $e^m$ such that $\xi_m \neq 0$. It is easy to see that we can write

$$\rho(a^m) = a^m \otimes 1 + (*) + 1 \otimes e^m$$

where the term $(*)$ is an A-linear combination of $1 \otimes e^{m'}$

such that $0 < |m'| < |m|$ . Applying the map $\rho$ to $\sum \xi_m a^m = 0$ and considering the A-coefficients of $1 \otimes e^m$ such that $|m| = n$, one obtains

$$\sum_{|m|=n} \xi_m \otimes e^m = 0 ,$$

a contradiction. Therefore the $a^m$ , $m \in [\Lambda]$, are $A^H$-linearly independent.

<u>We resume the proof of</u> 7.9.1: Since $A'$ is a free $A^H$-module, it is clearly a faithfully flat $A^H$-algebra. Since we have $A' \otimes_{A^H} A' \xrightarrow{\simeq} A' \otimes H$ , $a \otimes b \longmapsto (a \otimes 1)\rho(b)$ , it follows from (i) $\Longrightarrow$ (iii) of our proposition that the sequence

$$A' \xrightarrow{\partial^0} A' \otimes H \xrightarrow{\partial^1} A' \otimes H \otimes H$$

is exact. We have only to show $A = A'$. Since $A/A'$ is a right H-comodule, it suffices to see that the socle of $A/A'$ (as an H-comodule) is zero. Indeed if $a \in A$ is such that

$$\rho(a) - a \otimes 1 \in A' \otimes H ,$$

then since $\partial^1(\rho(a) - a \otimes 1) = 0$, there is an element $a' \in A'$ such that $\rho(a) - a \otimes 1 = \partial^0(a') = \rho(a') - a' \otimes 1$ . This means that $a = a' + (a-a') \in A' + A^H = A'$. Since H is irreducible, it follows that the socle of $A/A'$ is zero. Therefore $A = A'$. Done.

7.10. COROLLARY (to Theorem 7.1.1 = 7.8). <u>Let</u> $X = \mathrm{Spec}(A)$ <u>be an affine</u> k-<u>scheme on which an algebraic affine unipotent</u> k-<u>group scheme</u> $G$ <u>acts from the right. Let</u> $\rho: A \to A \otimes O(G)$ <u>be the associated comodule algebra structure map. Let</u> $B$ <u>be a subalgebra of</u> $A$ <u>such that</u> $\rho(B) \subseteq B \otimes O(G)$ <u>and put</u> $Y = \mathrm{Spec}(B)$. <u>If the action of</u> $G$ <u>on</u> $Y$ <u>is free and</u> $Y\tilde{\tilde{/}}G$ <u>is affine, then the action of</u> $G$ <u>on</u> $X$ <u>is free and</u> $X\tilde{\tilde{/}}G$ <u>is affine.</u>

<u>Proof</u>. It suffices to show that $H^1(G,B) = 0 \Longrightarrow H^1(G,A) = 0$. Let $K$ be a closed central subgroup scheme of $G$ such that $O(K)$ is generated by $P(O(K))$ as an algebra. Suppose that the assertion is true for $G\tilde{/}K$. If $H^1(G,B) = 0$, then the following groups all vanish in view of the exact sequence of 7.7.2 (and by the induction hypothesis):

$$H^1(G\tilde{/}K,B^K),\ H^2(G\tilde{/}K,B^K),\ H^1(K,B)^G,\ H^1(G\tilde{/}K,A^K) \text{ and } H^2(G\tilde{/}K,A^K)$$

and hence we have $H^1(G,A) \xrightarrow{\simeq} H^1(K,A)^G$. Let $x \in P(O(K))$. As was seen in the proof of 7.8, the 1-cocycle $1 \otimes x \in B \otimes O(K)$ is $G$-invariant. Since $H^1(K,B)^G = 0$, there is an element $b \in B$ such that $\rho(b) - b \otimes 1 = 1 \otimes x$. It follows from 7.9.1 that $H^1(K,A) = 0$. Hence $H^1(G,A) \simeq H^1(K,A)^G = 0$.

7.10.1. <u>Remark</u>. If $G$ acts freely on $Y$ and $Y\tilde{\tilde{/}}G$ is affine, then we have an isomorphism

$$A^G \otimes_{B^G} B \xrightarrow{\simeq} A.$$

This follows from the fact that the composite

$$A \otimes H \simeq A \otimes_B (B \otimes H) \simeq A \otimes_B (B \otimes_{B^G} B) \simeq A \otimes_{A^G} (A^G \otimes_{B^G} B)$$

$$\longrightarrow A \otimes_{A^G} A \simeq A \otimes H$$

is the identity and that $A$ is a faithfully flat $A^G$-algebra.

## 8. The underlying scheme of a unipotent algebraic group

*In this section (§8), after 8.1 the ground field k has a positive characteristic p. For affine k-group scheme G and a closed subgroup scheme H, we write G/H in place of $G\tilde{\tilde{/}}H$.*

8.0. Let G be an affine algebraic group scheme over a field k. If k is perfect, the following are equivalent — see DG-IV, §2, 3.9 and §4, 4.1(Thm. of Lazard); also Lazard's original proof in [3]:

(i) G is connected k-smooth unipotent.

(ii) G has a central series of closed subgroups with quotients isomorphic to $G_a$.

(iii) The underlying k-scheme of G is isomorphic to $\mathbb{A}^n = \mathrm{Spec}(k[X_1,\ldots,X_n])$ for some $n \geq 0$.

In this section we shall extend the above result to the case of non-perfect ground field. Hence we may and shall assume that the characteristic of k is $p>0$.

8.1. PROPOSITION. *For an algebraic k-group scheme G, the following are equivalent:*

(i) G *is connected, k-smooth and unipotent.*

(ii) G *has a central series of closed subgroups*

$$G = G_0 \supset G_1 \supset \ldots \supset G_n = (1)$$

*such that for all* $1 \leq i \leq n$, $G_{i-1}/G_i$ *is a* k-*form of* $(G_a)^{m(i)}$ *for some integer* $m(i)>0$,

*Proof.* (ii)⇒(i) is obvious. We prove (i)⇒(ii): Let $G$ be a unipotent algebraic k-group. Let $H_1=[G, G]$, $H_i=[G,H_{i-1}]$ for all $i>1$. We then obtain a standard central series $G \supset H_1 \supset \ldots \supset H_q=(1)$ in which all subgroups $H_i$ and their successive quotients $H_{i-1}/H_i$ are connected and k-smooth (see SGAD-VI$_B$, §7, for instance). We may therefore assume from the beginning that $G$ is commutative. Let $G^{p^\nu}$ denote the image of the $p^\nu$-th power operation $x \longmapsto x \cdots x$ ($p^\nu$ factors) inside the group $G$. We then obtain a central series

$$G \supset G^p \supset \cdots \supset G^{p^\nu} \supset \cdots \supset G^{p^N} = (1)$$

where again all $G^{p^\nu}$ and all successive quotients $G^{p^{\nu-1}}/G^{p^\nu}$ are connected k-smooth. Furthermore, these last quotients are killed by the p-th power operation $p \cdot \mathrm{id}$, so that by 1.6.1 their Verschiebungs are all zero. Therefore, by 1.7.1 each quotient $G^{p^{\nu-1}}/G^{p^\nu}$ is a k-form of $(G_a)^{m(\nu)}$, q.e.d.

8.2. Let $B$ be a commutative k-algebra. For integers $n \geq 0$, $m>0$, an $m \times m$ matrix $(\alpha_{ij})$ with entries in $k[F]$ and a column vector $Z=(z_i)_{i=1}^m$ with $z_i \in B$, we define a commutative B-algebra $B(n,(\alpha_{ij}),Z)$ as follows: Let $B[X_1,\ldots,X_m,Y_1,\ldots,Y_m]$ be the B-algebra of polynomials in $2m$ indeterminates $X_1,\ldots,X_m,Y_1,\ldots,Y_m$. Let $I$ be the ideal in $B[X_1,\ldots,X_m,Y_1,\ldots,Y_m]$ generated by

$$F^nY_i - \textstyle\sum_j \alpha_{ij}X_j - z_i, \quad i=1,\ldots,m,$$

(where one should recall that $F^rX_j=X_j^{p^r}$ (§1)). The quotient B-algebra $B[X_1,\ldots,Y_m]/I$ is denoted $B(n,(\alpha_{ij}),Z)$. The images of $X_i$, $Y_i$ in $B(n,(\alpha_{ij}),Z)$ are denoted by $x_i$, $y_i$ and called <u>the canonical generators of</u> $B(n,(\alpha_{ij}),Z)$. Thus

$$F^ny_i = \sum\alpha_{ij}x_j + z_i.$$

It is an easy exercise to show that

$$\{y_1^{e_1}\cdots y_m^{e_m}x_1^{f_1}\cdots x_m^{f_m} \mid 0 \leq e_i < p^n,\ 0 \leq f_i\}$$

form a B-basis of $B(n,(\alpha_{ij}),Z)$ (cf. The P-B-W theorem). Let $M=M(n,(\alpha_{ij}))$ be the left $k[F]$-module on generators $u_i$, $v_i$, $1\leq i\leq m$, defined by the set of $m$ relations

$$F^nv_i = \sum \alpha_{ij}u_j$$

as in 2.4. It is easy to see that the algebra map

$$\rho\colon B(n,(\alpha_{ij}),Z) \longrightarrow B(n,(\alpha_{ij}),Z) \otimes U(M)$$

$$x_i \longmapsto x_i \otimes 1 + 1 \otimes u_i$$

$$y_i \longmapsto y_i \otimes 1 + 1 \otimes v_i$$

is well-defined and makes $B(n,(\alpha_{ij}),Z)$ into a right $U(M)$-comodule algebra. By 7.9.1 the action of $\underline{\underline{D}}(M)$ on $\mathrm{Spec}(B(n,(\alpha_{ij}),Z))$ is free and the (fpqc)-quotient k-sheaf $\mathrm{Spec}(B(n,(\alpha_{ij}),Z))\tilde{/}\underline{\underline{D}}(M)$ is affine. Since

$$\{y_1^{e_1}\cdots y_m^{e_m}x_1^{f_1}\cdots x_m^{f_m} \mid 0 \leqq e_i < p^n,\ 0 \leqq f_i\}$$

are also a $B(n,(\alpha_{ij}),Z)^{\underline{\underline{D}}(M)}$-basis of $B(n,(\alpha_{ij}),Z)$ as is seen in the proof of 7.9.1, it follows that

$$B = B(n,(\alpha_{ij}),Z)^{\underline{\underline{D}}(M)}.$$

Conversely let $A$ be a commutative right $U(M)$-comodule algebra with $\rho\colon A \longrightarrow A \otimes U(M)$ the structure map. Suppose that the action of $\underline{\underline{D}}(M)$ on $\mathrm{Spec}(A)$ is free and that $\mathrm{Spec}(A)\tilde{/}\underline{\underline{D}}(M)$ is affine. Then by 7.9.1, there are $s_i$, $t_i \in A$, $i=1,\ldots,m$, such that $\rho(s_i)=s_i\otimes 1+1\otimes u_i$ and $\rho(t_i)=t_i\otimes 1+1\otimes v_i$. Then $c_i=F^n t_i-\sum\alpha_{ij}s_j$, $1\leqq i\leqq m$, belong to $A^{\underline{\underline{D}}(M)}$ clearly. Since

$$\{t_1^{e_1}\cdots t_m^{e_m}s_1^{f_1}\cdots s_m^{f_m} \mid 0 \leqq e_i < p^n,\ 0 \leqq f_i\}$$

form an $A^{\underline{D}(M)}$-basis of A by the proof of 7.9.1, it follows that the $A^{\underline{D}(M)}$-algebras A and $A^{\underline{D}(M)}(n,(\alpha_{ij}),(c_i))$ are canonically isomorphic. In fact the canonical isomorphism $A \simeq A^{\underline{D}(M)}(n,(\alpha_{ij}),(c_i))$ commutes with the action of $\underline{D}(M)$. This observation implies immediately in particular the following

8.2.1. LEMMA. *Let B be a commutative k-algebra. Let* $n, n' \geq 0$ *and* $m, m' > 0$ *be integers. Let* $(\alpha_{ij}) \in M_m(k[F])$ *and* $(\alpha_{ij}') \in M_{m'}(k[F])$ *be matrices with entries in* $k[F]$. *Suppose that the left* $k[F]$-*modules* $M(n,(\alpha_{ij}))$ *and* $M(n',(\alpha_{ij}'))$ *are isomorphic. Then for each vector* $Z=(z_i)_{i=1}^{m}$ *with* $z_i \in B$, *there is a vector* $Z'=(z_i')_{i=1}^{m'}$ *with* $z_i' \in B$ *such that the B-algebras* $B(n,(\alpha_{ij}),Z)$ *and* $B(n',(\alpha_{ij}'),Z')$ *are isomorphic.*

8.2.2. COROLLARY. *With the notations as above, suppose that* $(\alpha_{ij}(0)) \in GL_m(k)$, *where* $\alpha_{ij}(0)$ *denotes the constant term of* $\alpha_{ij}$. *Then the* $(\bar{k} \otimes B)$-*algebra* $\bar{k} \otimes B(n,(\alpha_{ij}),Z)$ *is isomorphic to the polynomial ring over* $\bar{k} \otimes B$ *in* m *indeterminates.*

*Proof*. Because $\bar{k} \otimes M(n,(\alpha_{ij})) \simeq \bar{k} \otimes M(0,(\alpha_{ij}))$ and $B(0,(\alpha_{ij}),Z)$ is a polynomial ring over B in m indeterminates.

8.3. *Definition*. A commutative k-algebra A is

said to be of type (*) if the following conditions are satisfied: For integers $N \geq 0$, $n(i) \geq 0$ and $m(i) > 0$ for $1 \leq i \leq N$, let $k[X_j^i, Y_j^i]$ be the polynomial k-algebra in the indeterminates $X_j^i$, $Y_j^i$, $1 \leq j \leq m(i)$, $1 \leq i \leq N$. There are elements $\alpha_{jk}^i \in k[F]$, $1 \leq j,k \leq m(i)$, $1 \leq i \leq N$ such that $(\alpha_{jk}^i(0))_{jk} \in GL_{m(i)}(k)$; and there are polynomials $z_j^i \in k[X_{j'}^{i'}, Y_{j'}^{i'}; 1 \leq j' \leq m(i'), 1 \leq i' < i]$, $1 \leq j \leq m(i)$, $1 \leq i \leq N$ such that $A$ is isomorphic to the quotient of $k[X_j^i, Y_j^i]$ by the ideal $I$ generated by

$$F^{n(i)} Y_j^i - \sum_k \alpha_{jk}^i X_k^i - z_j^i, \ 1 \leq j \leq m(i), \ 1 \leq i \leq N.$$

Let $B$ be a commutative k-algebra, $n \geq 0$ and $m > 0$ integers, $(\alpha_{ij}) \in M_m(k[F])$ a matrix such that $(\alpha_{ij}(0)) \in GL_m(k)$ and $(z_i)_{i=1}^m$ a column vector with $z_i \in B$. Then, if $B$ is of type (*), the k-algebra $B(n, (\alpha_{ij}), (z_i))$ is clearly of type (*). This remark and 8.2.2 prove by induction on $N$ the following

8.3.1. PROPOSITION. If $A$ is a commutative k-algebra of type (*), the $\bar{k}$-algebra $\bar{k} \otimes A$ is a polynomial algebra in finitely many indeterminates.

8.3.2. THEOREM. Let $G$ be an affine algebraic k-group scheme. The following are equivalent:

(i) $G$ is connected k-smooth unipotent.

(ii) G <u>has a central series of closed subgroup schemes with successive quotients isomorphic to</u> k-<u>forms of some products of finitely many copies of</u> $G_a$.

(iii) O(G) is a k-algebra of type (*).

<u>Proof</u>. (i)$\Longleftrightarrow$(ii) is 8.1 above. (iii)$\Longrightarrow$(i): Since $\bar{k} \otimes O(G) \simeq O(G \otimes \bar{k}) \simeq \bar{k}[X_1,\ldots,X_r]$, it follows from Theorem of Lazard that $G \otimes \bar{k}$ is connected $\bar{k}$-smooth unipotent. This proves (i). (ii)$\Longrightarrow$(iii): Let $G=G_0 \supset G_1 \supset \ldots \supset G_N=(1)$ be a central series of closed subgroups such that for each $1 \leq i \leq N$ $G_{i-1}/G_i$ is a k-form of $(G_a)^{m(i)}$ for some $m(i)>0$. By the induction hypothesis we may assume that $O(G/G_{N-1})$ is of type (*). By 2.5, there are an integer $n \geq 0$ and a matrix $(\alpha_{ij}) \in M_m(k[F])$ with $(\alpha_{ij}(0)) \in GL_m(k)$, where $m=m(N-1)$, such that

$$G_{N-1} \simeq \underline{D}(M(n,(\alpha_{ij}))).$$

Since the action of $G_{N-1}$ on G by the right translation is free and the quotient $G/G_{N-1}$ is affine, it follows from 8.2. that

$$O(G) \simeq O(G/G_{N-1})(n,(\alpha_{ij}),(z_i))$$

for some $z_1,\ldots,z_m \in O(G/G_{N-1})$. Since $O(G/G_{N-1})$ is of type (*), so is O(G). This proves (iii). Q.E.D.

8.4. We offer a remark supplementary to Lazard's Theorem 8.0:

8.4.1. PROPOSITION. <u>Let</u> G <u>be an affine</u> k-<u>group scheme such that</u> $\overline{G}\simeq \mathbb{A}^n$, <u>and let</u> H <u>be a</u> k-<u>closed normal subgroup scheme of</u> G <u>such that</u> $H\simeq G_a$. <u>Then</u>, $\overline{(G/H)}\simeq \mathbb{A}^{n-1}$.

<u>Proof</u>. Write $\overline{G}=\mathrm{Spec}\ k[t_1,\ldots,t_n]$ with indeterminates $t_1,\ldots,t_n$. By the Splitting Lemma (see Appendix), $\overline{G}\simeq \overline{(G/H)}\times \mathbb{A}^1$, whence we can write

$$k[t_1,\ldots,t_n] \simeq B \otimes k[u] \simeq B[u]$$

where $\overline{(G/H)}=\mathrm{Spec}\ B$ and $u$ is an indeterminate. Since we may clearly assume the k-algebra B to have a positive dimension, there exists an element $r\in B$ which is mapped onto a nonconstant polynomial $f(t_1,\ldots,t_n)$ under the isomorphism $B[u] \longrightarrow k[t_1,\ldots,t_n]$. Suppose as we may that f contains a term of minimal total degree (with respect to all $t_i$) involving $t_1$, the constant term being excepted. That is to say, suppose

$$f = a + bt_1^{\nu_1}t_2^{\nu_2}\cdots t_n^{\nu_n} + \cdots + ct_1^{\mu_1}\cdots t_n^{\mu_n} + \text{(terms of higher total degree, if any)},$$

where $\sum\nu_i=\cdots=\sum\mu_i=N(\text{say})$, $b\neq 0$ and $\nu_1>0$.

Let us take the automorphism of $k[t_1,\dots,t_n]$ defined by $t_1 \longmapsto t_1$, $t_2 \longmapsto t_2+\alpha_2 t_1,\dots,$ $t_n \longmapsto t_n+\alpha_n t_1$ with $\alpha_j \in k$ for $1 \leq j \leq n$. Under the automorphism $f$ is transformed to

$$g = a + (b\alpha_2^{\nu_2}\dots\alpha_n^{\nu_n}+\dots+c\alpha_2^{\mu_2}\dots\alpha_n^{\mu_n})t_1^N + \text{(mixed terms of total degree } N) + \text{(higher degree terms)}.$$

We may assume $k$ to be infinite, for otherwise the proposition is trivially true. Thus, it is possible to fix up $\alpha_2,\dots,\alpha_n \in k$ in such a manner that the coefficient of $t_1^N$ is nonzero. Consider now the homomorphism $B \longrightarrow k[t_1]$ obtained by composing the inclusion $B \hookrightarrow B[u]$, the isomorphism $B[u] \longrightarrow k[t_1,\dots,t_n]$, the automorphism of $k[t_1,\dots,t_n]$ fixed as above, and finally the homomorphism $k[t_1,\dots,t_n] \longrightarrow k[t_1]$ given by $t_1 \longmapsto t_1$, $t_j \longmapsto 0$ for all $j>1$. Under the homomorphism $B \longrightarrow k[t_1]$ just defined, the element $r$ is mapped onto $a+dt_1^N+$(higher degree terms in $t_1$, if any) with $d \neq 0$. We have thus obtained a nonconstant $k$-morphism $\mathbb{A}^1 = \operatorname{Spec} k[t_1] \longrightarrow \overline{(G/H)} = \operatorname{Spec} B$. Since $G/H$ is a unipotent algebraic $k$-group, this means that it is not $k$-wound by Tits' Theorem 4.3.1, whence follows that $\overline{(G/H)} \simeq \overline{N} \times \mathbb{A}^1$ for some unipotent $k$-group $N$. Then, $\overline{G} \simeq \overline{(G/H)} \times \mathbb{A}^1 \simeq \overline{N} \times \mathbb{A}^2$, or $k[t_1,\dots,t_n] \simeq C[u_1,u_2]$ where $\overline{N} = \operatorname{Spec} C$ and $u_1$, $u_2$ are indeterminates. One can simply repeat the above argument to find that $N$ is not $k$-wound. And so on. Thus, $\overline{(G/H)} \simeq \mathbb{A}^{n-1}$

is shown.

8.4.2. COROLLARY=[(iii)$\Rightarrow$(ii) of 8.0]. Note that our proof of this fact is essentially different from the one in DG-loc. cit. and is quite elementary.

8.4.3 COROLLARY. Let X be an affine k-scheme which is the underlying scheme of some k-group scheme. If $X\times\mathbb{A}^1 \simeq\mathbb{A}^n$, then $X\simeq\mathbb{A}^{n-1}$.

The proof is obvious. This corollary settles a (very) special case of the unsolved question: If an affine k-scheme X is such that $X\times\mathbb{A}^1\simeq\mathbb{A}^n$, is it then true that $X\simeq\mathbb{A}^{n-1}$?

## 9. The hyperalgebra of a unipotent group scheme

*In this section (§9), after 9.1 the ground field* $k$ *has a positive characteristic* $p$.

9.0. For a Hopf algebra $H$ over a field $k$, the dual Hopf algebra of $H$ will be denoted $H^0$ [13; page 122, §6. 2]. Let $G$ be an affine algebraic k-group scheme. The irreducible component of $O(G)^0$ containing 1 is called *the hyperalgebra of* $G$, denoted by hy(G) [16; 3.2.2]. The Lie algebra Lie(G) of $G$ is then isomorphic to P(hy(G)), the Lie algebra of primitive elements in hy(G) [16; Prop.3. 1.8].

Suppose that char(k) = 0. It is well-known then that $G$ is unipotent if and only if $G$ is connected and the Lie algebra Lie(G) = P(hy(G)) consists of nilpotent elements (the latter means that each element of Lie(G) acts on O(G) from the left as a locally nilpotent endomorphism). For a Hopf algebra $H$ we denote by $H^+$ the kernel of the counity $\varepsilon : H \longrightarrow k$. The following can be considered to be an analogy in the case of positive characteristic of the above result (cf. DG-IV, §2, 2.13):

THEOREM. *Suppose that* char(k) = $p > 0$ *and that* $G$ *is connected. Then* $G$ *is unipotent if and only if the ideal* $hy(G)^+$ *of* hy(G) *consists of nilpotent elements (in the sense of ring theory).*

9.1. Recall that a Hopf algebra H over k is called a hyperalgebra if it is cocommutative and irreducible [16; 1.3.5]. A hyperalgebra H is said to be of finite type if the set P(H) of primitive elements in H is finite dimensional.

9.1.1. LEMMA. Let H be a hyperalgebra of finite type (over a field of characteristic $p > 0$). Then H is a union of a directed family of finite dimensional sub-hyperalgebras*.

Proof. Let $x \in H$. Then x is contained in some finite dimensional sub-coalgebra C of H [13; page 46, Th.2.2.1]. Since C is pointed and irreducible, the dual algebra $C^*$ is local [13; page 160, Lem.8.0.2]. This means that the N-times iterated Frobenius map

$$F^N : (k,f^N) \otimes C^* \longrightarrow C^*, \ \lambda \otimes X \longmapsto \lambda X^{p^N}$$

(where $f^N : k \longrightarrow k, \lambda \longmapsto \lambda^{p^N}$) is identical with $\lambda \otimes X \longmapsto \lambda\langle X,1\rangle^{p^N}$ for sufficiently large N. Or, equivalently, the N-times iterated Verschiebung map

$$V^N : C \longrightarrow (k, f^N) \otimes C$$

(which is the dual coalgebra map of $F^N$) is equal to $c \longmapsto \varepsilon(c) \otimes 1$. (Notice that the map $V^N$ is denoted $T^N$ in [16; Prop. 1.9.1].) Let H' be the union of subcoalgebras D

* It follows that the finite dimensional sub-hyperalgebras of H form a directed set.

of $H$ such that $V^N(d) = \varepsilon(d) \otimes 1$ for all $d \in D$. Since the map $V^N : H \longrightarrow (k, f^N) \otimes H$ is a Hopf algebra map, it follows that $H'$ is a sub-hyperalgebra of $H$ which is killed by $V^N$. Since $x \in H'$, the assertion follows from the following:

9.1.2. LEMMA. *Let $C$ be a pointed irreducible cocommutative coalgebra* [13; page 157]. *Suppose that $P(C)$ (= the primitive elements of $C$ with respect to the unique group-like element) is finite dimensional. If the n-times iterated Verschiebung map $V^n : C \longrightarrow (k, f^n) \otimes C$ is trivial for some $n$, then $C$ is finite dimensional.*

*Proof*. Let $g_C$ be the unique group-like element in $C$. Let $M$ be the ideal $\{X \in C^* \mid \langle X, g_C\rangle = 0\}$ of $C^*$. Let $\{C_i\}$ be the coradical filtration of $C$ [13; page 185, §9.1]. Since $C_1 = kg_C + P(C)$ is finite dimensional, $C_i$ are all finite dimensional by [16; Prop.1.4.1]. We know that

$$C_i = \{x \in C \mid \langle M^{i+1}, x\rangle = 0\}$$

[13; page 220, Prop.11.0.5.]. Since $V^n : C \longrightarrow (k, f^n) \otimes C$ is $c \longmapsto \varepsilon(c) \otimes g_C$, it follows that

$$\langle X^{p^n}, x\rangle = 0 \text{ for all } X \in M \text{ and } x \in C.$$

Since $C^*$ is noetherian by [16; Prop.1.4.1], it follows

that the ideal $M$ of $C^*$ is nilpotent. If $N > 0$ is such that $M^{N+1} = 0$, then $C = C_N$ is finite dimensional.

9.2. PROPOSITION. *Let* $G$ *be an affine algebraic unipotent k-group scheme. Then all elements in* $hy(G)^+$ *are nilpotent.*

*Proof.* Let $x \in hy(G)^+$. Let $H$ be a finite dimensional sub-hyperalgebra containing $x$. Since $H \subseteq hy(G) \subseteq O(G)^0$, a natural Hopf algebra map

$$\pi : O(G) \longrightarrow H^*$$

is induced. Since $H$ is finite dimensional, $\pi$ is clearly surjective. In particular the Hopf algebra $H^*$ is irreducible, because so is $O(G)$. Hence $H$ is local by [13; Lem.8.0.2]. Therefore $x \in H^+$ is nilpotent (in the sense of ring theory).

9.3. PROPOSITION. *Let* $H$ *be a hyperalgebra of finite type. If* $H^+$ *consists of nilpotent elements, then the dual Hopf algebra* $H^0$ *is irreducible.*

*Proof.* Let $C$ be a finite dimensional subcoalgebra of $H^0$. Then the induced algebra map $H \longrightarrow C^*$ is clearly surjecteve. Hence there is a finite dimensional sub-hyperalgebra $H'$ of $H$ such that the restriction $H' \hookrightarrow H \longrightarrow C^*$ is surjective by 9.1.1. Since then $C \subseteq H'^0$, it follows that we have only to prove that $H'^0$ is irreducible. Thus we can assume that $H$ is finite dimensional from the beginning.

But, then, the ideal $H^+$ is easily seen to be nilpotent (see, e. g., [N. Jacobson, Lie Algebras, page 33, Th.1]). Hence $H^*$ is irreducible by [13; Lem.8.0.2].

9.4. Let $V$ be a k-vector space. A subspace $W$ of $V^* = \mathrm{Hom}_k(V, k)$ is said to be dense if the induced map $V \longrightarrow W^*$ is injective.

9.4.1. LEMMA. Let $G$ be an affine algebraic k-group scheme. If $G$ is connected, then $hy(G)$ is dense in $O(G)^*$.

Proof. Let I be the kernel of the canonical Hopf algebra map $O(G) \longrightarrow hy(G)^0$. Then $H = \mathrm{Spec}(O(G)/I)$ is a closed subgroup scheme of $G$ and $hy(G) \subseteq hy(H) \subseteq hy(G)$. It follows from [16; Prop.3.3.6] that $H$ is an open subgroup of $G$. Since $G$ is connected, $G = H$. Hence $I = 0$.

9.4.2. THEOREM. Let $G$ be a connected affine algebraic k-group scheme. Then, $G$ is unipotent if and only if the ideal $hy(G)^+$ of $hy(G)$ consists of nilpotent elements.

Proof. "Only if" part follows from 9.2. If $hy(G)^+$ consists of nilpotent elements, then $hy(G)^0$ is irreducible by 9.3. Since $O(G) \subseteq hy(G)^0$ by 9.4.1, it follows that $G$ is unipotent.

## APPENDIX

### Central extensions of affine group schemes

*In this Appendix, the ground field* k *is arbitrary, and all group schemes are affine.*

A.0. Our purpose here is to outline an elementary theory of central extensions of affine group schemes over an arbitrary ground field. Results summarized here are mostly known, but some are proven only over algebraically closed ground fields while others are known to be true in a more general context. (Cf. [5; Appendix], DG-III, §6 and [14; Chap. VII].) One of the two exact sequences below, A.8, is probably new, though it is modeled after an erroneously stated result by Miyanishi [5; Prop. 2, page 649].

A.1. A sequence $1 \to H \to E \to G \to 1$ of k-group schemes H, E, G is said to be k-*exact* or, plainly, *exact* if (i) all arrows represent k-homomorphisms, (ii) for any k-algebra R the sequence $1 \to H(R) \to E(R) \to G(R)$ of abstract groups is exact, and (iii) $E \to G$ is a faithfully flat epimorphism. The sequence, or often E itself, is referred to as an *extension of* G *by* H. If $1 \to H \to E \to G \to 1$ is k-exact and if the image of H is contained in the center of G, then we say the sequence is *central* k-*exact* or is a *central* extension of $G$ by $H$. We also say that $E$ is a central extension of $G$ by $H$, by a slight abuse of language.

A.2. Let $1 \to H \to E \to G \to 1$ be k-exact. The sequence is said to be k-split if there is a k-homomorphism $G \to E$ such that $(G \to E \to G) = id_G$. The sequence is geometrically k-split, by definition, if there exists a k-morphism $\overline{G} \to \overline{E}$ such that $(\overline{G} \to \overline{E} \to \overline{G}) = id_{\overline{G}}$. Then, obviously, $\overline{E} \simeq \overline{H} \times G$.

SPLITTING LEMMA. *An exact sequence* $1 \to H \to E \to G \to 1$ *of k-group schemes is geometrically k-split whenever* $H$ *is k-isomorphic to* $G_a$.

This is implied by DG-III, §4, 6.6. Within the classical framework, this was first proved by Rosenlicht [9; Th. 1, p.99]. Let us, however, prove it in a few lines by another method:

*Proof*. Let $\overline{H} = \text{Spec } k[T]$, $\overline{E} = \text{Spec } A$ and $\overline{G} = \text{Spec } B$. Since $H \simeq G_a$ acts freely on $\overline{E}$ with the quotient $\overline{G}$, we have $A \simeq k[T] \otimes B \simeq B[T]$ by virtue of [4; Lemma 2, p. 403], and the ring homomorphism $A \to B$ given by $T \mapsto 0$ evidently splits the sequence geometrically over $K$.

A.3. *Remarks*. The consequences of this Lemma are important. Firstly, we immediately derive the fact that the underlying scheme of a unipotent k-group of dimension $n$ is k-isomorphic to $\mathbb{A}^n$ if $k$ is *perfect*, or more generally if the group is k-<u>solvable</u> (see DG-IV, §4, 4.1 and [9; Cor. 2, p.101]). Secondly, this lemma is the key to make Serre's induction argument [14; VII, No. 10, pp. 175-177] valid over non-algebraically closed fields, thereby establishing Chevalley's theorem over a <u>perfect</u> field k: To wit, every commutative unipotent algebraic k-group is k-isogenous to a unique (up to order) product of Witt vector groups. (Analyzing Serre's argument [14; <u>loc. cit</u>], one can see directly that what fails the argument in case k is merely assumed perfect is the inability to affirm, in that set-up, that the extensions $0 \to G_a \to E \to W_n \to 0$ over k correspond to $H^2_{reg}(W_n, G_a)_s$. This one can now ascertain, possessing a regular cross section $W_n \to E$ defined over k, thanks to the Lemma. This gives a proof of Chevalley's theorem without resort to the theory of Dieudonné modules, as done in DG-V, §3, 6.11.)

A.4. If a central extension $0 \to H \to E \to G \to 1$ of k-group schemes is geometrically k-split, the group law on E is determined by 2-<u>cocycles</u> $G \times G \to H$ in the well-known manner. To wit, the group multiplication on $\bar{E} = \bar{G} \times \bar{H}$ is given by

$$(g_1,h_1)(g_2,h_2) = (g_1g_2,\ h_1+h_2+\gamma(g_1,g_2))$$

for $g_1,g_2 \in G(R)$, $h_1,h_2 \in H(R)$, where $\gamma: G \times G \to H$ is a k-morphism satisfying

$$\gamma(g_1,g_2) + \gamma(g_1g_2,g_3) = \gamma(g_1,g_2g_3) + \gamma(g_2,g_3)$$

for all $g_1,g_2,g_3 \in G(R)$. Conversely, given a 2-cocycle $\gamma\colon G \times G \to H$, one can construct a central extension $0 \to H \to E \to G \to 1$ by defining a group law on $\bar{G} \times \bar{H}$ in the above fashion by making use of $\gamma$. <u>The extension thus obtained will be denoted by</u> $G \times_\gamma H$.

A.5. Let $1 \to H \to E_1 \to G \to 1$, $1 \to H \to E_2 \to G \to 1$ be extensions of $G$ by $H$. We say that these extensions are <u>equivalent</u> if there is a k-homomorphism $E_1 \to E_2$ making the diagram

$$\begin{array}{ccccccccc} 1 & \to & H & \to & E_1 & \to & G & \to & 1 \\ & & \| & & \downarrow & & \| & & \\ 1 & \to & H & \to & E_2 & \to & G & \to & 1 \end{array}$$

commutative. The set of equivalence classes of extensions of $G$ by $H$ is denoted by $\mathrm{Ext}(G,H)$. In case $H$ is commutative, we consider the set of equivalence classes of central extensions of $G$ by $H$ and denote it by $\mathrm{Ext}_{cent}(G,H)$, which is a subset of $\mathrm{Ext}(G,H)$. If in addition $G$ is commutative, the set of all equivalence classes of commutative extensions is represented as $\mathrm{Ext}_{com}(G,H)$.

A.6. Let $G$ be a k-group scheme, $H$ a commutative k-group scheme. Let $0 \to H \to E \to G \to 1$ be a central extension, and suppose given a k-homomorphism $\phi: G' \to G$ of k-group schemes. Then, one can construct a central extension $0 \to H \to E' \to G' \to 1$ unique up to equivalence subject to the commutativity of the diagram

$$\begin{array}{ccccccccc} 0 & \to & H & \to & E' & \to & G' & \to & 1 \\ & & \| & & \downarrow & & \downarrow\phi & & \\ 0 & \to & H & \to & E & \to & G & \to & 1. \end{array}$$

We write $\phi^*E$ in place of $E'$. In a like manner, for a given k-homomorphism $\psi; H \longrightarrow H'$, one can construct $\psi_*E$ uniquely up to equivalence subject to the commutativity of the diagram

$$\begin{array}{ccccccccc} 0 & \to & H & \to & E & \to & G & \to & 1 \\ & & \psi\downarrow & & \downarrow & & \| & & \\ 0 & \to & H' & \to & \psi_*E & \to & G & \to & 1. \end{array}$$

Possessing $\phi^*E$ and $\psi_*E$, one can proceed to introduce a structure of additive group on the set $\mathrm{Ext}_{\mathrm{cent}}(G,H)$ in the usual fashion, and $\mathrm{Ext}_{\mathrm{cent}}(G,H)$ becomes a right $\mathrm{End}_{k\text{-}gr}(G)$-and left $\mathrm{End}_{k\text{-}gr}(H)$-bimodule. The constructions and verifications pertaining to the foregoing are found in [14; Chap. VII, §1], [5; Appendix], DG-III, §6 and SGAD-III, $VI_A$,

XVII-App.1., though with various degrees of generality.

A.7. LEMMA. *Let* $\phi: G \to H$ *be a* k-*homomorphism of* k-*group schemes* G, H. *If the subgroup* $\phi(G(R))$ *is normal* (*resp. central*) *in* $H(R)$ *for every* $R \in \boxed{Alg}_k$, *then the image* $\phi(G)$ *is a* k-*closed normal* (*resp. central*) *subgroup of* H.

In fact, the image $\phi(G)$ is the (fpqc)-sheafication of the k-group functor $R \longmapsto \phi(G(R))$, so that if $x \in \phi(G)(R)$ and $h \in H(R)$ then there is a faithfully flat R-algebra R' such that $x' \in \phi(G(R'))$, where $x \longmapsto x'$ under $\phi(G)(R) \to \phi(G)(R')$. By assumption, $(h')^{-1}x'h' \in \phi(G(R')) \subseteq \phi(G)(R')$, and then, by (fpqc)-descent, one gets $h^{-1}xh \in \phi(G)(R)$. Because k is a field, $\phi(G)$ is clearly k-closed. Similar argument for the centrality of $\phi(G)$ when each $\phi(G(R))$ is central.

A.8. THEOREM. *Let* $0 \to H \overset{\tau}{\to} E \overset{\pi}{\to} G \to 1$ *be a central extension of* k-*group schemes*, H *being commutative*. *For every commutative* k-*group scheme* A, *consider the sequence of additive groups*

$$0 \to \mathrm{Hom}_{k\text{-}gr}(G,A) \to \mathrm{Hom}_{k\text{-}gr}(E,A) \to \mathrm{Hom}_{k\text{-}gr}(H,A)$$

$$\overset{\gamma}{\to} \mathrm{Ext}_{cent}(G,A) \overset{\pi^*}{\to} \mathrm{Ext}_{cent}(E,A) \overset{\tau^*}{\to} \mathrm{Ext}_{cent}(H,A) \quad (1)$$

*where* $\gamma$ *sends* $\phi \in \mathrm{Hom}_{k\text{-}gr}(H,A)$ *to the extension class of*

$\phi_*E$. *Then:*

(i) *The sequence* (1) *is exact except possibly at* $\mathrm{Ext}_{\mathrm{cent}}(E,A)$, *where only* $\tau^*\pi^* = 0$ *holds in general.*

(ii) *Assume that* G, H *and* A *are* k-*smooth and that* $\mathrm{Hom}_{k_s\text{-gr}}(H,A) = \{0\}$. *Then, the sequence* (1) *is exact throughout.*

*Proof.* (i) We shall make quick verification of exactness at each spot, leaving out all routine procedures.

a) The exactness at $\mathrm{Hom}(G,A)$ and at $\mathrm{Hom}(E,A)$ is trivial.

b) Next, at $\mathrm{Hom}_{k\text{-gr}}(H,A)$, suppose $\phi \in \mathrm{Hom}(H,A)$ factors through $\tau$ so that $\phi = \psi\tau$ with $\psi: E \to A$. Then define a k-homomorphism $E \to A \times E$ by $x \longmapsto (-\psi x, x)$ for all $x \in E(R)$ and compose it with the canonical homomorphism $A \times E \to \phi_*E$. The composed homomorphism $E \to \phi_*E$ vanishes on the closed subgroup scheme H and thereby gives a k-homomorphism $G \to \phi_*E$. It is immediate that $(G \to \phi_*E \to G) = \mathrm{id}_G$, so $\phi_*E$ represents the zero in $\mathrm{Ext}_{\mathrm{cent}}(G,A)$. Suppose now that $\phi_*E$ is trivial. Then we have a canonical projection $\rho: \phi_*E \simeq A \times G \to A$ with the property $(A \to \phi_*E \to A) = \mathrm{id}_A$. Thus, in the diagram

$$\begin{array}{ccccccccc}
0 & \to & H & \xrightarrow{\tau} & E & \xrightarrow{\pi} & G & \to & 1 \\
 & & \phi\downarrow & & \downarrow\lambda & & \| & & \\
0 & \to & A & \underset{\rho}{\overset{\tau'}{\rightleftarrows}} & \phi_*E & \to & G & \to & 1
\end{array}$$

we have $(\rho\lambda)\tau = \rho(\tau'\phi) = \phi$, whence $\phi$ factors through $\tau$ <u>via</u> $\rho\lambda$.

c) At $\mathrm{Ext}_{\mathrm{cent}}(G,A)$, first it is clear that the composition $\mathrm{Hom}(H,A) \to \mathrm{Ext}_{\mathrm{cent}}(G,A) \to \mathrm{Ext}_{\mathrm{cent}}(E,A)$ is zero, as $\phi \longmapsto \phi_*E \longmapsto \pi^*(\phi_*E) = \phi_*(\pi^*E)$ and $\pi^*E$ is trivial. Next let $0 \to A \to X \to G \to 1$ be a central extension such that $\pi^*X$ is trivial, and let us show that $X = \phi_*E$ for an appropriate $\phi: H \to A$. So, consider the diagram

$$\begin{array}{ccccccccc}
0 & \to & A & \to & X & \to & G & \to & 1 \\
 & & \| & & \uparrow & & \uparrow\pi & & \\
0 & \to & A & \to & \pi^*X & \to & E & \to & 1 \\
 & & & \nwarrow^{-\phi} & \uparrow & & \uparrow\tau & & \\
 & & & & H & = & H & &
\end{array}$$

in which $\pi^*X \simeq A \times E$ admits a projection to $A$. Set $-\phi$ to be the said projection preceded by the k-monomorphism $H \to \pi^*X$, as shown above. Then, under the identification of $\pi^*X$ with $A \times E$, the monomorphism $H \to \pi^*X$ is given by $h \in H(R) \longmapsto (-\phi h, \tau h)$ and $X$ is the quotient of $\pi^*X = A \times E$ by the image of $H$. This shows that $X = \phi_*E$.

d) Finally, at $\mathrm{Ext}_{\mathrm{cent}}(E,A)$, it is obvious that $\tau^*\pi^* = (\pi\tau)^* = 0$.

(ii) Assume now $\mathrm{Hom}_{k_s\text{-}gr}(H,A) = \{0\}$ and $G,H,A$ (and hence $E$) are k-smooth. Let $0 \to A \to X \to E \to 1$ be a central extension such that $\tau^*X$ splits. Then we have a k-homomorphism $\psi : H \to X$ obtained by composing the section on $H$ with the canonical homomorphism $\tau^*X \to X$, and clearly $\rho\psi = \tau$ where $\rho: X \to E$ is the given faithfully flat homomorphism. Let $R$ be a $k_s$-algebra, $x \in X(k_s)$ and $h \in H(R)$. Then, as $H$ is central in $E$, we have $\rho[x(\psi h)x^{-1}(\psi h)^{-1}] = (\rho x)(\tau h)(\rho x)^{-1}(\tau h)^{-1} = e \in E(R)$, so that we may write $x(\psi h)x^{-1} = (\alpha h)(\psi h)$ with $\alpha h \in A(R)$. One verifies right away the relation $\alpha(h + h') = \alpha h + \alpha h'$ and thereby obtains a $k_s$-homomorphism $\alpha: H_{k_s} \to A_{k_s}$ which is constant by assumption. Consequently, $\psi(H(R))$ commutes with $X(k_s)$. But, A and E being k-smooth, X, too, is k-smooth so that $X(k_s)$ is dense in X, whence follows that $\psi(H(R))$ is central in $X(R)$. Now, since the image functor $R \longmapsto \psi(H(R))$ is central in X, the image itself is central by virtue of A.6. Let $Y = X/\mathrm{Im}\psi$. It is routine to check that Y is a central extension of G by A such that $\pi^*Y = X$, as desired.

Q.E.D.

A.9. THEOREM. <u>Let</u> $0 \to B \xrightarrow{\tau} C \xrightarrow{\pi} A \to 0$ <u>be a commutative extension of commutative k-group schemes and let</u> G <u>be a k-group scheme. Then, the following sequence of additive groups</u>

$$0 \to \mathrm{Hom}_{k\text{-}gr}(G,B) \to \mathrm{Hom}_{k\text{-}gr}(G,C) \to \mathrm{Hom}_{k\text{-}gr}(G,A)$$

$$\xrightarrow{\gamma} \mathrm{Ext}_{cent}(G,B) \xrightarrow{\tau_*} \mathrm{Ext}_{cent}(G,C) \xrightarrow{\pi_*} \mathrm{Ext}_{cent}(G,A) \qquad (2)$$

<u>is exact, where</u> $\gamma$ <u>sends</u> $\phi \in \mathrm{Hom}_{k\text{-}gr}(G,A)$ <u>to the extension class of</u> $\phi^*C$.

Proof of this theorem is omitted, as it is routine, similar to that of A.8. Besides, the theorem is essentially known — cf. SGAD, <u>loc</u>. <u>cit</u>. and DG, <u>loc</u>. <u>cit</u>.

A.10. <u>Example</u>. Consider $0 \to \alpha_p \to G_a \xrightarrow{F} G_a \to 0$. Coupling the three group schemes with $G_a$, one obtains two complexes as in A.8 and A.9. The second is exact. As for the first, the part

$$\mathrm{Ext}_{cent}(G_a,G_a) \xrightarrow{F^*} \mathrm{Ext}_{cent}(G_a,G_a) \to \mathrm{Ext}_{cent}(\alpha_p,G_a)$$

is not exact. Indeed, note that $\mathrm{Ext}_{cent}(\alpha_p,G_a) \cong k$, while $\mathrm{Ext}_{cent}(G_a,G_a)$ is a free left $k[F]$-module with a countable basis $u_0, u_1, \cdots, u_n, \cdots$ such that $F^*(u_i) = Fu_i$ for all $0 \leq i < \infty$. ( See 3.6.1.) The non-exactness of the sequence above is therefore evident.

## INDEX OF TERMINOLOGY

# TABLE OF NOTATION

## REFERENCES


DG: M. Demazure-P. Gabriel, "Groupes Algébriques, tome I," 1970, Masson & Cie, Éditeur-Paris; North-Holland Publ. Co.-Amsterdam.

SGAD: "Schémas en groupes, I, II, III, dirigés par M. Demazure et A. Grothendieck," 1970, Springer-Verlag—Berlin, Heidelberg, New York.

FGA: A. Grothendieck, "Fondements de la Géométrie Algébriques — extraits du séminaire Bourbaki,1957-62," Secrétariat Mathématique — Paris.

EGA: A. Grothendieck, "Eléments de Géométrie Algébrique," 1960ff, Publ. Math. I.H.E.S. — Paris.

[1] N. Bourbaki, "Algèbre, Chapitres 6 et 7," 1964, Hermann — Paris.

[2] T. Kambayashi, A note on groups in a category, Math. Zeitschr. 93(1966), 289-293.

[3] M. Lazard, Sur nilpotence de certains groupes algébriques, C. R. Acad. Sci. Paris 241(1955), 1687-1689.

[4] M. Miyanishi, On the vanishing of the Demazure cohomologies and the existence of quotient preschemes, J. Math. Kyoto Univ. 11(1971), 399-414.

[5] __________, Une caractérisation d'un groupe algébrique simplement connexe, Ill. J Math. 16(1972), 639-650.

[6] __________, Some remarks on the polynomial rings, to appear in Osaka J. Math.

[7] __________, "Introduction à la théorie des sites et son application à la construction des préschemas quotients," No.47, Publ. du Séminaire de Math. Sup., 1971, Les Presses de l'Univ. de Montréal-Montréal.

[8] M. Rosenlicht, Some rationality questions on algebraic groups, Annali Mat. Pura Appl. (4) 43(1957), 25-50.

[9] __________, Questions of rationality for solvable algebraic groups over nonperfect fields, Annali Mat. Pura Appl. (4) 61(1963), 97-120.

[10] __________, Automorphisms of function fields, Trans. Amer. Math. Soc. 79 (1955), 1-11.

[11] P. Russel, Forms of the affine line and its additive group, Pacific J. Math. 32(1970), 527-539.

[12] P. Samuel, "On unique factorization domain," Lectures on Math. and Physics 28(1967), Tata Inst. Fund. Research-Bombay.

[13] M. Sweedler, "Hopf Algebras," 1969, W. A. Benjamin & Co. — New York.

[14] J. -P. Serre, "Groupes algébriques et corps de classes," 1959, Hermann - Paris.

[15] J. Tits, "Lectures on Algebraic Groups," 1966/67, Dept. of Math., Yale Univ. - New Haven.

[16] M. Takeuchi, Tangent coalgebras and hyperalgebras, I, to appear in J. Math. Soc. Japan.

[17] A. Grothendieck, Sur quelques points d'algèbre homologique, Tôhoku Math. J. 9(1957), 119-221.

Vol. 247: Lectures on Operator Algebras. Tulane University Ring and Operator Theory Year, 1970–1971. Volume II. XI, 786 pages. 1972. DM 40,–

Vol. 248: Lectures on the Applications of Sheaves to Ring Theory. Tulane University Ring and Operator Theory Year, 1970–1971. Volume III. VIII, 315 pages. 1971. DM 26,–

Vol. 249: Symposium on Algebraic Topology. Edited by P. J. Hilton. VII, 111 pages. 1971. DM 16,–

Vol. 250: B. Jónsson, Topics in Universal Algebra. VI, 220 pages. 1972. DM 20,–

Vol. 251: The Theory of Arithmetic Functions. Edited by A. A. Gioia and D. L. Goldsmith VI, 287 pages. 1972. DM 24,–

Vol. 252: D. A. Stone, Stratified Polyhedra. IX, 193 pages. 1972. DM 18,–

Vol. 253: V. Komkov, Optimal Control Theory for the Damping of Vibrations of Simple Elastic Systems. V, 240 pages. 1972. DM 20,–

Vol. 254: C. U. Jensen, Les Foncteurs Dérivés de lim et leurs Applications en Théorie des Modules. V, 103 pages. 1972. DM 16,–

Vol. 255: Conference in Mathematical Logic – London '70. Edited by W. Hodges. VIII, 351 pages. 1972. DM 26,–

Vol. 256: C. A. Berenstein and M. A. Dostal, Analytically Uniform Spaces and their Applications to Convolution Equations. VII, 130 pages. 1972. DM 16,–

Vol. 257: R. B. Holmes, A Course on Optimization and Best Approximation. VIII, 233 pages. 1972. DM 20,–

Vol. 258: Séminaire de Probabilités VI. Edited by P. A. Meyer. VI, 253 pages. 1972. DM 22,–

Vol. 259: N. Moulis, Structures de Fredholm sur les Variétés Hilbertiennes. V, 123 pages. 1972. DM 16,–

Vol. 260: R. Godement and H. Jacquet, Zeta Functions of Simple Algebras. IX, 188 pages. 1972. DM 18,–

Vol. 261: A. Guichardet, Symmetric Hilbert Spaces and Related Topics. V, 197 pages. 1972. DM 18,–

Vol. 262: H. G. Zimmer, Computational Problems, Methods, and Results in Algebraic Number Theory. V, 103 pages. 1972. DM 16,–

Vol. 263: T. Parthasarathy, Selection Theorems and their Applications. VII, 101 pages. 1972. DM 16,–

Vol. 264: W. Messing, The Crystals Associated to Barsotti-Tate Groups: With Applications to Abelian Schemes. III, 190 pages. 1972. DM 18,–

Vol. 265: N. Saavedra Rivano, Catégories Tannakiennes. II, 418 pages. 1972. DM 26,–

Vol. 266: Conference on Harmonic Analysis. Edited by D. Gulick and R. L. Lipsman. VI, 323 pages. 1972. DM 24,–

Vol. 267: Numerische Lösung nichtlinearer partieller Differential- und Integro-Differentialgleichungen. Herausgegeben von R. Ansorge und W. Törnig. VI, 339 Seiten. 1972. DM 26,–

Vol. 268: C. G. Simader, On Dirichlet's Boundary Value Problem. IV, 238 pages. 1972. DM 20,–

Vol. 269: Théorie des Topos et Cohomologie Etale des Schémas. (SGA 4). Dirigé par M. Artin, A. Grothendieck et J. L. Verdier. XIX, 525 pages. 1972. DM 50,–

Vol. 270: Théorie des Topos et Cohomologie Etale des Schémas. Tome 2. (SGA 4). Dirigé par M. Artin, A. Grothendieck et J. L. Verdier. V, 418 pages. 1972. DM 50,–

Vol. 271: J. P. May, The Geometry of Iterated Loop Spaces. IX, 175 pages. 1972. DM 18,–

Vol. 272: K. R. Parthasarathy and K. Schmidt, Positive Definite Kernels, Continuous Tensor Products, and Central Limit Theorems of Probability Theory. VI, 107 pages. 1972. DM 16,–

Vol. 273: U. Seip, Kompakt erzeugte Vektorräume und Analysis. IX, 119 Seiten. 1972. DM 16,–

Vol. 274: Toposes, Algebraic Geometry and Logic. Edited by. F. W. Lawvere. VI, 189 pages. 1972. DM 18,–

Vol. 275: Séminaire Pierre Lelong (Analyse) Année 1970–1971. VI, 181 pages. 1972. DM 18,–

Vol. 276: A. Borel, Représentations de Groupes Localement Compacts. V, 98 pages. 1972. DM 16,–

Vol. 277: Séminaire Banach. Edité par C. Houzel. VII, 229 pages. 1972. DM 20,–

Vol. 278: H. Jacquet, Automorphic Forms on GL(2). Part II. XIII, 142 pages. 1972. DM 16,–

Vol. 279: R. Bott, S. Gitler and I. M. James, Lectures on Algebraic and Differential Topology. V, 174 pages. 1972. DM 18,–

Vol. 280: Conference on the Theory of Ordinary and Partial Differential Equations. Edited by W. N. Everitt and B. D. Sleeman. XV, 367 pages. 1972. DM 26,–

Vol. 281: Coherence in Categories. Edited by S. Mac Lane. VII, 235 pages. 1972. DM 20,–

Vol. 282: W. Klingenberg und P. Flaschel, Riemannsche Hilbertmannigfaltigkeiten. Periodische Geodätische. VII, 211 Seiten. 1972. DM 20,–

Vol. 283: L. Illusie, Complexe Cotangent et Déformations II. VII, 304 pages. 1972. DM 24,–

Vol. 284: P. A. Meyer, Martingales and Stochastic Integrals I. VI, 89 pages. 1972. DM 16,–

Vol. 285: P. de la Harpe, Classical Banach-Lie Algebras and Banach-Lie Groups of Operators in Hilbert Space. III, 160 pages. 1972. DM 16,–

Vol. 286: S. Murakami, On Automorphisms of Siegel Domains. V, 95 pages. 1972. DM 16,–

Vol. 287: Hyperfunctions and Pseudo-Differential Equations. Edited by H. Komatsu. VII, 529 pages. 1973. DM 36,–

Vol. 288: Groupes de Monodromie en Géométrie Algébrique. (SGA 7 I). Dirigé par A. Grothendieck. IX, 523 pages. 1972. DM 50,–

Vol. 289: B. Fuglede, Finely Harmonic Functions. III, 188. 1972. DM 18,–

Vol. 290: D. B. Zagier, Equivariant Pontrjagin Classes and Applications to Orbit Spaces. IX, 130 pages. 1972. DM 16,–

Vol. 291: P. Orlik, Seifert Manifolds. VIII, 155 pages. 1972. DM 16,–

Vol. 292: W. D. Wallis, A. P. Street and J. S. Wallis, Combinatorics; Room Squares, Sum-Free Sets, Hadamard Matrices. V, 508 pages. 1972. DM 50,–

Vol. 293: R. A. DeVore, The Approximation of Continuous Functions by Positive Linear Operators. VIII, 289 pages. 1972. DM 24,–

Vol. 294: Stability of Stochastic Dynamical Systems. Edited by R. F. Curtain. IX, 332 pages. 1972. DM 26,–

Vol. 295: C. Dellacherie, Ensembles Analytiques, Capacités, Mesures de Hausdorff. XII, 123 pages. 1972. DM 16,–

Vol. 296: Probability and Information Theory II. Edited by M. Behara, K. Krickeberg and J. Wolfowitz. V, 223 pages. 1973. DM 20,–

Vol. 297: J. Garnett, Analytic Capacity and Measure. IV, 138 pages. 1972. DM 16,–

Vol. 298: Proceedings of the Second Conference on Compact Transformation Groups. Part 1. XIII, 453 pages. 1972. DM 32,–

Vol. 299: Proceedings of the Second Conference on Compact Transformation Groups. Part 2. XIV, 327 pages. 1972. DM 26,–

Vol. 300: P. Eymard, Moyennes Invariantes et Représentations Unitaires. II. 113 pages. 1972. DM 16,–

Vol. 301: F. Pittnauer, Vorlesungen über asymptotische Reihen. VI, 186 Seiten. 1972. DM 18,–

Vol. 302: M. Demazure, Lectures on p-Divisible Groups. V, 98 pages. 1972. DM 16,–

Vol. 303: Graph Theory and Applications. Edited by Y. Alavi, D. R. Lick and A. T. White. IX, 329 pages. 1972. DM 26,–

Vol. 304: A. K. Bousfield and D. M. Kan, Homotopy Limits, Completions and Localizations. V, 348 pages. 1972. DM 26,–

Vol. 305: Théorie des Topos et Cohomologie Etale des Schémas. Tome 3. (SGA 4). Dirigé par M. Artin, A. Grothendieck et J. L. Verdier. VI, 640 pages. 1973. DM 50,–

Vol. 306: H. Luckhardt, Extensional Gödel Functional Interpretation. VI, 161 pages. 1973. DM 18,–

Vol. 307: J. L. Bretagnolle, S. D. Chatterji et P.-A. Meyer, Ecole d'été de Probabilités: Processus Stochastiques. VI, 198 pages. 1973. DM 20,–

Vol. 308: D. Knutson, λ-Rings and the Representation Theory of the Symmetric Group. IV, 203 pages. 1973. DM 20,–

Vol. 309: D. H. Sattinger, Topics in Stability and Bifurcation Theory. VI, 190 pages. 1973. DM 18,–